THE BOOK OF THE PISTOL AND REVOLVER

Originally published in 1921

ISBN: 978-1-940001-25-8

Published by Sportsman's Vintage Press
2014

Wheelocks and Snaphaunce (Italian)

PLATE 1.

TO

DIANA

THE BOOK OF THE
PISTOL & REVOLVER

By

CAPTAIN HUGH B. C. POLLARD

Author of "A Busy Time in Mexico," etc.

CONTENTS

APPENDICES

THE BOOK OF THE PISTOL AND REVOLVER

CHAPTER I

INTRODUCTION

FROM the earliest period of the invention of firearms, mankind has desired a short, easily portable, and easily concealed means of defence, and to this common desire we owe the pistol, the revolver, the "automatic," and all their kindred. The pistol—by which term I include all pistols, revolvers, automatics, and repeating or single-shot weapons meant for use in one hand—was essentially designed as a weapon for quick use at close quarters. It was a weapon of offence or defence with a limited range, particularly suitable for horsemen, and endowing their possessor with an advantage over enemies more skilled in the use of the *arme blanche* or physically superior to him.

In tracing the history and development of the pistol throughout the ages, we find that in the main it has always been the weapon of the mounted gentleman and companion to the sword. In the hey-day of the duelling period the disastrous cult of the hired bravo and bully imperilled the foundations of the code of honour, and as a revulsion towards a more equable method of settling differences, pistol duelling, with its infinitely greater equalization of risk, superseded the use of the small sword.

It is easy to infer that even then degrees of skill in the use of pistols were equally well marked as those of prowess in fencing; but as a matter of historical fact, while fencing was always the jealously guarded privilege of the gentry and military classes, the use of a pair of pistols was a necessary factor in the education of every householder, gentle or simple, during the period of the sixteenth, seventeenth, and eighteenth centuries. It is safe to say that no household and few travelling carriages did not boast of the armament of a pair of pistols as a necessary defence against highwaymen, rogues, and masterless men.

During the early period of crude match-lock and arquebus, petronel and wheel-lock, body armour and plate of proof still survived, and it was only with the development of firearms and the consequent increased range of combat, together with the difficulty of constructing armour sufficiently strong to turn a ball and yet light enough to be worn without undue fatigue, that pistols began to come to their own.

In the days when firearms were rare, and their use confined (owing to their costliness) to the nobles, we find that the early pistols and dags were modelled along the lines of weapons built to serve two purposes—those of a pistol, and a club or mace for use at close quarters—and they were usually, owing to their rarity, made the vehicles for some of the finest applied art of the period.

During the Renaissance the craftsman was at the zenith of artistic achievement, and many an artist silversmith was also an armourer of no mean order.

Weapons were a necessity to every man, and upon them was lavished a wealth of loving skill and creative art that has lived to enrich the museums of our modern nations. In utility the firearms of the past were for the most part as accurate and efficient as could be achieved, considering the limitations of the tools at their disposal and the explosives with which they were acquainted. Love for their handicraft inspired the armourers; and many passages in the Memoirs of Benvenuto Cellini, a Florentine artist and silversmith of 1500-1571, proclaim how little has changed the amateur's love for fine weapons.

With the invention of the percussion system an enormous impetus was given to the development of the pistol, and this invention was the point which made the revolver anything beyond a cumbrous and inefficient repeating weapon.

Revolving firearms of all kinds have been known since the dawn of the age of gunpowder, but the necessity for embodying cumbrous flash-pans, and the danger of other priming charges of loose powder lying in close proximity to the explosion and gas escape of the first discharge, had always made them ungraceful and dangerous weapons.

The percussion system changed that, and simultaneously in all countries the improved revolver made its appearance, but it remained for the citizens of the United States, then engaged upon the dangerous and arduous task of enlarging their territories and establishing their social code, to develop the Revolver.

To them and the French we owe the succeeding improvements and the applications of breech-loading which have since taken place, leading us step by step to that wonderful weapon, the modern revolver, and its ultimate successor, the automatic pistol.

Despite the claims of inventors of European origin, there is no doubt that the revolver in all its best applications is the outcome of American ingenuity and workmanship, and one may well term it their national weapon. To this day there lingers in the murderous Republic of Mexico the tradition of the terrible revolver-armed Texans of the early days of the Lone Star State, and even now, in out of the way parts, your skilled revolver shot earns the whispered sobriquet of "El Tejano" ("the Texan") a man with whom quarrels are to be avoided!

As a factor in adventure and romance the pistol and revolver are all paramount, and to establish a library of books wherein all mention of these weapons was excluded would, I take it, exclude some two-thirds or more of our known standard works.

The love of weapons is deep-rooted in the hearts of manly men, and skill at arms throughout the ages has been extolled as becoming to the character and necessary to the development of all good citizens, so among the mass we always find a chosen few to whom the use and knowledge of weapons is an all-absorbing hobby, and whose ideal is the collection of representative, and even the rare and unusual, types of weapons in their historical development.

The average schoolboy has a wholesome interest in firearms, and at nearly every school occur a few rare youthful spirits who, in the pursuit of their hobby, defy the masters and the powers that be, revelling in the secret possession of firearms. This is a trait condemned by all who realize the danger of loosely handled firearms, but in extenuation one must plead that accidents with weapons in the hands of those who understand and love them are very, very few, whereas our journals constantly record deaths from the "did not know it was loaded" class—the bulk of fools who have not wit enough to master the first simple facts of use of arms.

Should parents be troubled by a son who loves pistols, I would advise them, first, to see that the weapons be of good quality, and not likely to burst, or go off accidentally, and secondly, to substitute, if possible, a gun or rifle for the pocket-arm, for it must be remembered that a pistol muzzle moves in a much smaller radius than the longer barrelled arms, and that with them accidents are more easily avoided and game far more easily killed. I do not believe in discouraging a lad in the practice of arms, although I would teach him care and economy by means of a Spartan severity in the matter of ammunition.

The reason for this is that when every cartridge is of value and supply exceedingly limited, your true boy learns perforce to perfect his shooting, does not indulge in risky shots likely to wound rather than kill game, and also learns a deal of woodcraft and the habits of his quarry.

I myself started a career of shooting with a flint-lock horse pistol "borrowed" from a trophy of arms. Powder was dreadfully scarce, and narrow-minded shopmen would not supply a boy of eight with it. Urged by these economic conditions my sole source of supply was rook-rifle cartridges, filched from a splendidly careless uncle; the powder from these blended with match-heads, and, later, a home-made meal powder of my own (God knows what risks I ran!) was all I had; newspaper wads, and, for shot, nicely graduated pebbles from the gravelled paths. With these difficulties to contend with I allowanced myself one shot a day, and, as a rule, devoted all the holiday morning to that shot. In the summer holidays, when young rabbits abounded, better game for a child to stalk were never found. The old pistol would kill (perhaps!) at seven to ten yards, and to get within this range needed infinite patience, skill, and the development of a faculty for taking cover. I learnt more in that blissful summer than ever I could set out in cold print; learnt tricks of the wild, secrets of wind and track, that have stood me in good stead in pursuit of game in such divers places as Mexico or the North African Hinterland.

Of course it was all very wicked, but stolen pleasures are often sweetest, and it conveys the solemn moral—one shot, one kill.

To write a true book of all pistols requires a lifetime of study, travel, and research, and would fill many volumes, a delight only to the collector, and of little use to the ordinary man. My endeavour is to cover in this volume a brief summary of pistols in their wide variations, and to view them from the standpoint of the modern amateur of arms.

Exactitude in the matter of dates is often beyond one, but I have done the best to approximate as closely as possible.

I plead then in advance for my shortcomings, my disregard for all but the simplest tables, and my neglect of many quite charming firearms, whose individual peculiarities would interest us for more than a moment.

Criticism must find a place, for without it a book would be little more than a well-bound catalogue, and so I pray that those who disagree with me will remember that they also believe in their theories.

CHAPTER II

THE DAWN OF POWDER

THE origin of gunpowder is veiled in the mists of antiquity, and it is not known when it first came into use as a propelling agent. As a combustible and as a firework it is early mentioned, and its probable earliest use in war was as a rocket or bomb.

Gentoos' code of laws, which was formed at about the same period as Moses, contains a reference. "The magistrate shall not make war with any deceitful machine, or with poisoned weapons, or with cannon and guns, or any kind of firearms," but it is quite probable that "cannon" means "artillery," which may be taken as meaning any kind of siege engine, and firearms might for the matter of that have been "fire arrows."

In 1267 Roger Bacon alludes to gunpowder, and gives a recipe for its preparation, which he is assumed to have got from Marcus Graecus, who lived about the end of the eighth century, and describes a very sound method of making a large rocket.

Artillery were certainly in use at the Siege of Granada in 1325, and a year later the Gonfaloniere and Council of Twelve of Florence ordered brass cannon and iron shot for the defence of their city. Their use spread rapidly, and Edward III. of England considerably astonished the Scotch by using "crakys of war," in 1327.

The eastern origin of gunpowder is probably due to the fact that only in dry, hot countries is saltpetre found in its natural state, and it is certain that in 1626 the East India Company started to import saltpetre for the Government, though Continental Powers had to lay down special "farms" for its production.

The early powder was a simple mixture of saltpetre, charcoal, and sulphur, ground up together in the proportion of seventy-five of saltpetre, fifteen of charcoal, ten of sulphur. This was known as "meal powder," and was used both as a charge and for priming.

During the reign of Elizabeth, the process of granulating or corning was discovered, by which the mixture is broken up into small grains. This process has been in use ever since, as it vastly improves the powder and gives it greater strength. The reason for this is that it allows the flame of ignition to penetrate more speedily between the grains than it could throughout a solid mass of meal powder.

Modern black powder has the grains glazed by the admixture of graphite in the corning mill, a process which partially waterproofs them. From time to time the nations varied the composition of their powder, and even powder for the private use of sportsmen was of very variable strength.

A century ago it was quite customary for sportsmen to test their powder in a little device known as an "éprouvette." This was a little flint-lock pistol action machine, which, in place of blowing out a bullet, blew round a cover attached to a graduated disc that revolved against a spring. The position of the disc after discharge gave the reputed strength of the powder.

There is extant an account of a duel between Charles James Fox and Adams. They were at Brook's, when Fox made some remark about bad Government powder. Adams, who was responsible, took the matter up and challenged Fox.

They met, and Adams fired, but Fox refused to, saying: "I'll be damned if I do, for I have no quarrel."

They shook hands after the encounter, and Fox, who was slightly wounded, said:

"Adams, you'd have killed me if it had not been Government powder."

Nowadays, black powder has been superseded by smokeless compounds, but it is still on sale in country districts, and made on a commercial scale for blasting—anyway, it has made history for six hundred years; and whatever may be claimed for Berthold Schwartz, it is clear that gunpowder was not invented by a German.

The actual date of the invention of gunpowder is lost, and so also is the date of origin of the first pistol; but one can safely assume that certain specimens of Chinese match-lock pistols represent the first form of the pistol. These are for the most part small, immensely heavy castings, either round or octagon in section, and of exceptionally small bore in proportion to their thickness. The shape is usually that of a small trunnionless cannon, and they were unprovided either with stock or any system of attachment to a handle other than simple binding to a stick. No lock work or serpentin to hold the match is provided, and discharge was accomplished by merely pressing the lighted end of a length of match or rope upon the touch-hole. Some more modern ones are provided with a tang or extension, which folded back over the barrel and touch-hole, and was sufficiently springy to hold away the lighted match until the whole was grasped in the hand, when it was required to fire the piece. Later the usual lock and serpentin were adopted.

In Europe hand firearms were at first practically baby cannon, of which the human body formed the trail and a forked crutch stuck in the ground the support, and none of these can truly be regarded as pistols, as all needed two hands to manage—one to hold the weapon, and the other to apply the match. The next, and most important, development was the Match-lock, which was popular from its inception to the close of the last century, where a very perfect form of this weapon was used by the tribes of the Indian North-West Frontier.

The match-lock in its simplest form consisted of a barrel fastened to a wooden stock by bands. In the rear of the barrel, and in a slot in the stock, was pivoted a curved lever or serpentin, the lower part of which—that is to say, the portion below the axis—was much heavier than the top. This weighted portion served to keep the jaws of the serpentin, which held a lighted match, away from the powder in the flash-pan. The match, a length of rope soaked in a saltpetre solution, was held so that when the lower end of the serpentin, which served also as trigger, was pulled up toward the stock, the glowing head descended into the loose powder of the pan, causing discharge. In later models the match was carried coiled upon the barrel, and the serpentin held but a small length of match, which one first lit from the main match, and then, by further pressure, plunged into the pan. The source of this invention of the serpentin, so called from its snake-like outline and fire-holding jaws, was probably due to the application of the releasing mechanism of the cross-bow to the hand firearm. The flash-pan of match-locks was always protected by a crudely fitted lid arranged to swing laterally, and designed to guard against accidental discharge and the effects of damp.

Although no European specimens survive, it hardly seems likely that short-barrelled match-locks capable of use in one hand (pistols, in a word) were not in use, and there is no doubt that the knightly outcry against the use of firearms (and very loud it was) created a demand for a weapon easily usable by a mounted man, and it must always be realized that pistols are essentially a cavalryman's weapons.

Hand guns were in use between 1375 and 1400, and the match-lock, usually known as arquebus, about 1476, being first established among the Swiss, but represented in 1518 before our King Henry VIII. at the Field of the Cloth of Gold. The Yeomen of the Guard, a corps established by

Henry VII., were armed half with bows and half with arquebuses, and in Hans Burgman's picture of the "Triumph of Maximilian" can be seen the complete equipment of the arquebusier of the sixteenth century.

The pistol, so called either from the fact that its calibre corresponded with the "pistole," the coin, or from its reputed invention at Pistoia, in Italy, was common in Germany in 1512, and was adopted by the French cavalry in 1550. One Caminelleo Vitelli, of Pistoia, is reputed to have invented it about 1540, but its undoubted previous popularity disposes of his claim.

About the period of 1510-1520 there seems to have been an outbreak of enthusiasm among inventors, due possibly to the stimulus of the wars between France and the States of Italy, and about this time occurred the invention of a new system of ignition which was about to reverse all previous conceptions of the limited utility of firearms.

The European match-lock differs from the ordinary Oriental model in that the serpentin points toward the stock, as does the cock of a wheel-lock. In all Oriental arms the serpentin is reversed and occupies a position similar to the usual hammer of a firearm, pointing from the stock toward the barrel.

CHAPTER III

THE WHEEL-LOCK

UP till 1515 all firearms had been handicapped by the necessity of having a burning match attached. This was undesirable and dangerous for many reasons, notably that the flame and smoke of the match betrayed the position of the gunner, a point that rendered the match-lock unsuitable as a sporting weapon or upon such military enterprises as night attacks. After all, the process had been cumbrous, for the musketeer had first to load with coarse powder, a tow wad, a bullet, and another wad, and then to "prime his piece"—that is to say, to fill the flash-pan into which the serpentin dipped with fine powder; then with flint and steel to light his match and carefully nurse the glowing spark into a lambent red-hot head; then had the little piece in the serpentin to be lit, and so at last the weapon was ready to be fired. Attempts had probably been made to attach a tinder-box to the side of a gun, but all had failed. Of early attempts at providing self-ignition the "Monks Gun" in the Dresden Museum is one of the best-known examples. This is similar in action to the pop-gun of our nursery days. Alongside the barrel below the touch-hole (which is at the side) is placed a metal box in which moves backwards and forwards a serrated plunger, the extension of which forms a handle. From the top of the plunger case rises a spring in the shape of a serpentin, which presses a piece of flint down upon the serrated plunger. In use the weapon was held in one hand and the plunger worked rapidly backwards and forwards like a bicycle pump, so that the friction of flint and steel produced a spark in the vicinity of the touch-hole and fired the charge. This weapon may be regarded as the legitimate ancestor of the wheel-lock.

The wheel-lock proper was invented in 1515 at Nuremberg, and was a revolutionary departure from all previous methods of firearm construction. The principle was fairly simple, consisting as it does of utilizing the effect of a notched steel disc, revolving at high speed against a stationary flint or fragment of iron pyrites to produce a stream of sparks in the flash-pan.

In wheel-locks the hammer lies forward of the lock, and, unlike all other firearms, its jaws point towards the butt. This is, in a way, an incorrect statement, for the hammer of the wheel-lock is, properly speaking, a serpentin, but instead of holding a match the jaws are made of stouter metal, and hold a piece of flint or iron pyrites.

This is pressed down by a spring at the other end of the serpentin into a pan of powder which communicates by a touch-hole with the charge in the barrel.

This pan has its bottom formed of a steel disc, on the rim of which are chased serrations or channels, and this disc or wheel is outside on the face of the lock, and from it springs the name, "wheel-lock."

This disc was connected by a short link of chain to a very strong spring inside the lock, and was wound up or spanned with a key to a point, where a pin on the trigger lever carried a projecting stud into a hole in the flat surface of the disc, and locked it in the cocked position.

The flash-pan was provided with a sliding lid, which was connected with the internal mechanism of the lock in such a manner as to slide back, exposing the priming powder and allowing the stone in the serpentin to drop into contact with the steel disc when it revolved. This occurred when the trigger was pulled and the stud drawn back into the lock, freeing the disc to revolve against the spring's compression.

The lock was good, but its construction delicate, and at that period exceedingly expensive. So far as it went it was effective, and as a military weapon was sometimes combined with a match-lock device in case the other lock went out of action.

To fire the piece the flash-pan was primed, the wheel wound or spanned, and the serpentin lowered on the flash-pan lid. When the trigger was pulled, the wheel was free to revolve and driven by the spring spun round at high speed, generating sufficient streams of sparks to fire the priming. In later models an improvement was added in the shape of a pin release to the pan cover. On pressing this projecting pin the pan cover flew back, uncovering the pan for a fresh priming charge to be inserted.

The speedy popularization and perfection of

this device gave a great impetus to the manufacture of pistols proper, for now weapons could be concealed about the person and no cumbrous match was necessary. It also proved an ideal weapon for the hunter, as a steady shot could be obtained, and was soon manufactured in pistol and demi-pistol sizes to suit the needs of mounted men.

The first form of wheel-lock pistol was the "dag," or "tacke." These were short-barrelled, heavy, clumsy weapons, usually of large bore, and with clumsy butts terminating in large knobs or balls—probably with the idea of serving the dual purpose of pistol and club or mace. Practically all sixteenth and seventeenth century pistols were so mounted, and were splendidly carved, embossed, and inlaid. The tradition of the knob butt still lingers, and Tartar and Mongol pistols of comparatively recent manufacture still adhere to the style, having a knob trigger and a heavy ivory ball butt.

In the Wallace Collection are many splendid specimens of wheel-lock pistols, or dags, and they are very often made by goldsmiths and jewellers, as well as by armourers, for in those times the craftsman in metals was frequently a general all-round mechanic, besides being a skilled and painstaking artist. The conception, design, and workmanship of Italian weapons of the wheel-lock period is often such as to render them gems of artistic production; and to this we owe many of the beautiful specimens that have been preserved to the present day.

Sometimes one comes across baby wheel-locks —mere toys, an inch or two in length, of no practical value as weapons, yet perfect models of the larger pistols. These are supposed to be the work of the young apprentices of gun and silver smiths, who showed their skill with their tools by making little models of their masters' work, exactly as do modern apprentices in the engineering works of to-day.

Many dags were bell-mouthed, and some excellent double-barrelled, both side by side—and under and over—specimens may be found. The tradition of the bell-mouthed, short-barrelled and short-stocked pistol continued for a long time in Spain, where quite recently similar weapons made for breech - loading shot - gun cartridges were manufactured. The same taste in weapons is noticeable in the North African littoral, where a bell-mouthed, or rather trumpet-mouthed, pistol, with a short stock, often carved or inlaid, and frequently of miniature gun-stock shape, was a favourite weapon among the Arabs of the Tripoli, Algiers, and Morocco. (A similar weapon is in the possession of the author, having been brought back from Tripoli by Mr. Allen Ostler, the War Correspondent for the *Daily Express* at the time of the Italian campaign. It was then in use, and was brought back loaded.)

Neither the dag in its original form, nor the later straight-barrelled pieces, were wonderfully accurate; some were provided with crude sights, but the majority, and certainly the bell-mouthed pattern, were intended for use at close quarters, and were loaded with several slugs to form a scattering charge, rather than a single bullet.

The wheel-lock was soon the best military arm,

and was adopted by all regiments rich enough to afford them, although poorer battalions still retained the match-lock. During the reigns of Mary and Elizabeth, the official weapons embraced carabines, petronels, which were really long-barrelled, long-stocked wheel-lock pistols, and also a hybrid firearm called the "dragon," whence we get the name "dragoons." The "dragon" was a brass barrelled semi-pistol semi-carbine, with a fanciful dragon mouth to the barrel. The dragoons were meant to operate as mounted infantry.

The pistol of the Armada period was the long-barrelled, wheel-lock petronel, and it is worthy of note that Queen Elizabeth encouraged the manufacture of gunpowder and weapons in the country, although there is no doubt that at that time, and for a good deal later, Continental weapons were superior in material and finish. During the Civil War, both Cavaliers and Roundheads used pistols as part of the equipment of a mounted man. There was no standard bore or length, but in Charles II.'s time it was insisted upon that cavalry pistols should not be less than fourteen inches long over all. The average weapon was the petronel, whose long stock, almost in the same line as the barrel, frequently approximated in shape rather to the gunstock than the easily grasped pistol.

In England wheel-locks were manufactured in small quantities, and in perfectly plain material, without decoration or finish. The stocks were made of light, soft woods, and it is seldom that they have lasted to the present day. The majority of wheel-locks in existence are the more decorated pieces of German or Italian manufacture, but English pieces are rare—rarer than the armour of the Civil War period, for their wood rotted and they were thrown away.

The name "petronel" comes either from *poitrine-al*, the French word *poitrine* meaning "chest," and indicating the mode of firing, the weapon being butted up against the breast; or from the Spanish word *pedregal*, "stone," in reference to the stone or flint held in the lock. Either explanation is good, but neither certain.

In 1630 the flint-lock was introduced, and this was but little used in England until much later, but it is safe to assume that early flint-locks were being used, particularly by rich Cavaliers, during the Civil War period; though for all practical purposes it may be taken that the wheel-lock period continued from the time of Henry VIII., whose Yeomen of the Guard carried wheel-lock "dagges" in 1519 right down through Elizabethan times—the period of James I., Charles I., and the Civil Wars. They were not officially superseded until the reign of William III. and the beginning of the eighteenth century.

CHAPTER IV

THE FLINT-LOCK

THE type of lock known as the Flint-lock was popular for more than two centuries, during which time it never underwent any substantial change in design. It was applied to every kind of firearm, from pocket pistols to cannon ; and was the standard military system from the time of Charles II. to Waterloo.

The majority of collectable pistols belong to this period, and by far the greater number of them to the period of 1775-1830. To those collectors who have greater depth of purse, the earlier types are still accessible ; but to the ordinary person the flint-lock period is the one that offers the greatest variety of interest for the smallest outlay.

There are two existing accounts of the origin of the flint-lock. One claims that it was invented by the Spanish or Portuguese irregulars, or "Miguelitos"—hence the French name, *Lock à Miguelet"*; and the other tradition gives it Dutch origin, claiming that it was invented by the Dutch chicken thieves, "Snap-haans," who could not afford wheel-locks, and found the match-lock, with its flaming match, unsuitable for their nocturnal trade. There is no proof of either of these legends being correct, and I have another theory to advance.

From 1542 the Portuguese had been trading with Japan. Existent at this time in Japan was a small form of flint-lock, not applied to pistols or fire-arms at all, for these were unknown, but used as a portable tinder-box. These flint-lock tinder-boxes are usually egg-shaped, and about the size of a small hen's egg. They remind one of the modern "Easter egg," and open into two sections, exposing the flint-lock with a small receptacle for tinder. The trigger is usually an external stud. I advance the theory that these handy little devices would naturally appeal to the traveller of the period, and that some were brought back by traders or sailors to Portugal, where the principle was adapted to fire-arms. Its spread from the Peninsula under the name of *Lock à Miguelet* would naturally follow.

Many of these egg tinder-boxes are works of art, signed by well-known Japanese artists and metal workers. I have had no means of locating the earliest examples, but should any collecter possess signed or dated specimens, I would be grateful for a note of the information sent to me in care of my publishers.

The earlier form of flint-lock was the type known as "Snaphaunce," and it is intermediate between the real flint-lock and the wheel-lock. As it was only made during a short period, it is the rarest of all forms of pistol, and the value of specimens is rapidly increasing, few being left outside museums or large private collections.

The Wallace Collection and the Birmingham Art Gallery have beautiful collections of snap-haunces, mainly of Italian workmanship; and it is seldom that snaphaunces by European makers other than Italians are met with, though they were made in England as early as 1585.

Their difference from the flint-lock is unmistakable, for they retain the sliding pan cover of the wheel-lock, and the steel or frizzen is not combined with the cover, as it is in the flint-lock, but independent.

The internal mechanism of the lock is similar to that of the wheel-lock, for the same system is employed to uncover the pan. The snaphaunce principle is sometimes met with in Tartar and Caucasian pistols of later date, but the finest specimens are the Italian weapons by Lazzaro Comminazo, and Geo. Maria Aquisti.

The standard flint-lock came in about 1630, and was popular for some two hundred years. The principle of the flint-lock is simplicity itself. A flake of flint held in the jaws of a pivoted lever, now called the "hammer" but then called the "cock" is driven by a spring against an upright extension of the cover of the flash-pan, called the "frizzen." This when struck by the flint is thrown back, uncovering the priming powder, upon which the sparks resultant from the collision of flint and cover fall. The cumbrous winding up or "spanning" of the wheel-lock was thus done away with, together with all other businesses of uncovering the flash-pan, and setting down the serpentin.

There are two other forms of flint-lock that deserve notice, and these are the types used on Spanish and Arab firearms, in which the whole of the lock mechanism is external, and of two types. In one the mainspring pushes up, in the other down. The trigger action of both these types is distinctive, and evidently adapted from the wheel-lock.

In the wheel-lock the rotating disc is held cocked by a pin that passes into a hole in the disc from inside the lock plate. In the Spanish lock this pin is retained, and serves to support an extension of the hammer. Pressing the trigger draws the pin back by means of a bell-crank lever, and the hammer is brought down by the pressure of the spring. All Spanish and Oriental locks of this type have serrated steel frizzens dovetailed into the soft iron of the pan cover.

European weapons have almost invariably plain steel pan covers. These serrations are distinctively Oriental. The Spanish have always been ultra-conservative, and the Moorish tradition still continues. In modern Spain one can find cheap breech-loading fowling-pieces with the same pin release to the hammer as was used on the earliest firearms.

Flint-lock pistols fall into three main groups, according to the purpose for which they were designed, but the line of demarcation between the groups is frequently crossed by weapons designed to suit general utility purposes, hybrids between the two main types.

The great divisions of the period are: (1) Horse or holster pistols, which embraces all large pistols; military and naval models; (2) duelling pistols; and (3) pocket pistols. Into one or other of these categories nearly all muzzle-loading pistols can be placed.

During the latter part of the seventeenth century, in Charles II.'s reign, it was ordered that the standard military pistol should not be less than sixteen inches in length over all. By 1700 the pistol began to grow less cumbersome, and the average size was about twelve inches for officers' weapons, though the rank and file still retained the cumbrous and crude earlier type. The notable weapon of the Queen Anne period is the type known as the "cannon barrel." This variety of pistol is distinguished by the shape of the barrel, which is outwardly turned to resemble an old-fashioned cannon. The barrel is screwed to a breech-block, which contains a tapering-powder chamber of less diameter than the bore of the piece, and the whole is designed for loading from the breech with a well-fitting ball, thus obviating all danger of the bullet rolling out, the barrel being more or less chambered or relieved at the rear to accommodate it.

The cannon barrel pistol was made in a variety of sizes and in several types, but the earlier specimens are usually from ten to twelve inches over all, about twenty-two bore, and have a distinctive lock. The action is fitted in a box behind the breech of the barrel ; the hammer is *at the side*, and the frizzen spring lies underneath the pan, curving round to a point midway between the pan and the hammer. A specimen in the author's possession carries on its silver mounts the hall mark for 1705.

The second type of cannon barrel is the later model, probably about 1720, and has an ordinary side lock attached to the wooden butt, but not part of the barrel. The wood fore end terminates abruptly as soon as it is clear of the lock. The lock is held to the butt by two screws passing through a metal strap on the opposite side.

Models of this description are usually found in pocket sizes, and were made for concealment in the large "salt box" pockets of the costumes of the period.

The third and most common model of cannon barrel has an inside box lock extension to the breech, but the hammer is not at the side, but vertical, and in the line of sight. The frizzen and flash-pan are on the top of the barrel, and it is similar to most of the flint-lock pocket pistols in having no possible line of sight along the barrel. The dates of this type may be anywhere between 1740 and 1800.

The rare Scotch or Highland pistols were often of this cannon barrel type, and some were equipped with a ring butt to serve as a key for unscrewing the barrel of the companion pistol. These Scotch pistols were made during the seventeenth and eighteenth century to match Highland costume, and were made with metal stocks, either plain or engraved, in steel, or occasionally in silver. The locks may be either outside hammer or enclosed, and in some the Spanish type of lock is preserved. The triggers are of the knob type, similar to the Russian Cossack pistols. At present they command an exorbitant price, and modern imitations or "conversions" are not unknown.

All types of cannon barrel pistols were highly ornamented and inlaid, and most butts terminate in a silver or brass mask. They were essentially gentlemen's weapons and expensive to make, but contemporary with them ran the standard military pistols and pairs of horse pistols of the period.

The early flint-lock horse pistols had substantial metal knob butts, and the stock was straighter than the conventional "pistol grip." The knob butt served a double object in securing balance and serving as a club or mace for use at close quarters.

The style of butt is termed by some "fishtail," as the knob bulges out just as a fish's tail splays outward. From the metal knob run up two slender curved wings of metal inlaid into the wood of the stock, these serving to strengthen it at its weakest part.

The origin of this type is probably Italian, but all good European arms of the period, other than cannon barrels, used this type of butt.

The barrels were round, made of thin steel, and thickening towards the breech. In some cases the shape turns to octagon for the breech end, but completely octagon barrels are a later type.

The muzzles were slightly "swamped" or belled outward. This must not be confused with bell-mouthed pieces, but means that if a straight-edge is laid along the barrel from breech to muzzle the line of the barrel shows us a graceful curve. This principle gave better balance owing to the weight of metal at the muzzle, and was easy to sight. It was only employed on good class weapons.

The distinguishing marks of the early eighteenth-century horse pistols are thin barrels, knob butts, and bent lock plates—that is to say, the lower edge of the lock plate is not straight, but dips towards the stock. The section of the hammer is *oval*, a hammered forging, not as if the hammer had been cut out of a flat piece of steel, and there are seldom any sights.

As the art of manufacture advanced, barrels got heavier and more accurately welded. Grace of design and a wealth of inlay and ornament distinguished the Italian and French weapons, and at the middle of the eighteenth century a gentleman's pair of pistols were objects of art. Sights, with the exception of a bead for a foresight, were unusual ; but the graceful design of the stocks, the balance and the "lay" of the pistols, ensured an automatic accuracy, merely pointing them at an object bringing them on with sufficient accuracy to hit your man instinctively at distances up to twenty paces.

The pistols of the period were made in styles of finish suitable to the social order of the age : exquisite works of art for the nobility, plain but graceful and reliable weapons for the gentry, and coarse, cumbrous, but efficient equipment for the middle and lower classes.

Pistols were made as a rule in pairs, but cases did not come in until the days of duelling pistols, as most weapons were sold with leather holsters for attachment to the saddle.

The period of 1750 to 1775 develops the pistol from the ornate cannon barrel and decorated horse pistol toward a weapon of greater weight and solidity, capable of vastly greater accuracy. The tendency is best described as abandoning shot-gun lines of construction for the heavier and more deadly rifle.

After 1780 duelling with small swords went out of favour, and affairs of honour were settled with the pistol rather than the blade. Stimulated by this, makers turned their attention to the production of a heavy, accurate pistol, warranted to hit a halfpenny at twenty paces, and pistol design was so soon influenced by the development of this superior weapon that all other types of pistol speedily followed the prevailing mode.

The luxurious mountings, chiselling, and inlay gave place to a plain serviceable weapon of first-class workmanship and magnificent mechanical and ballistic efficiency. The graceful round barrel of light metal yielded to the heavier octagonal form. Bright silver fittings were replaced by bronzed or blued steel that would not glitter in the sun and attract an adversary's aim, and locks of sure and certain action were of greater value than graceful ornament.

The English makers won a reputation surpassed by none for the plain excellence of workmanship, and solid reliable design suited their natural expression as makers of weapons rather than makers of objects of art, and it was in their hands that the British duelling pistol reached its apogee as the most accurate and reliable flint-lock pistol ever produced. They were plain, but exquisite in craftsmanship, shooting power, and balance—a triumph of the gunsmith's art.

The names of Joseph Manton, Mortimer, Durs Egg, Wogdon, Nock, Blanch, and Rigby, were known and honoured all the world over wherever pistols were used or needed.

The English duelling pistol in its first period

retained some of the graceful lines of its lighter forerunners, but later examples sacrifice grace to sterner needs.

The typical duelling pistol is of half-inch bore or less, has a nine or ten-inch octagon barrel of smooth bore, and a plain, light trigger pull. The stocks are usually of walnut, and free of ornament. The barrel is secured to the stock by flat holding-down bolts, and the lock strap—the piece of ornamental metal on the opposite side to the lock—is done away with, though the lock is held to the stock by two screws termed "side nails." The ramrod is carried under the barrel, and is horn, *not metal*, tipped, and the wood of the stock extends to the muzzle.

The same type of pistol endured till superseded or converted to the percussion system, but certain small changes came in. One of these was the introduction of the "detent," a moving intercepting lever in the lock, whose function was to prevent a hammer slipping while being cocked, and firing the charge. This it accomplished by guiding the scear into the half-cock notch upon the tumbler, and so arresting the blow of the hammer before it struck the frizzen. Next in importance came an improvement to the frizzen action.

The blow of the hammer has to force the frizzen back round a pivot, and against a spring on the outside of the lock plate. This spring has two functions to perform. One is to give the frizzen sufficient resistance to strike sparks from the flint in the hammer jaws; the other to keep the pan cover section of the frizzen down tight on the pan to prevent accidental loss or damping of the priming powder. By about 1790 it was the custom to place a small wheel built in with the frizzen spring at the point of friction between frizzen stem and spring. This wheel acted as a roller, and eliminated the slow friction, with the result that the pan opened much more quickly, and the priming received many more sparks, ensuring a quicker and more certain ignition.

Another detail of interest is the extension of the trigger-guard toward the barrel. This is usually more or less decorated, and of fancy design; the earlier weapons up to, say, 1790, have an acorn pattern, possibly of Jacobite origin. Towards 1800 and thence onward this develops into an unmistakable pineapple, possibly a design started by the West India trade.

Military and naval weapons were distinguished by the crudeness and simplicity of their manufacture, the later models having a metal ramrod attached by a ringed stirrup to the stock to prevent its loss while reloading. They, as well as many other types of pistol, are often provided with long belt hangers at the side for carrying in an accessible position while in action. With the exception of a few very late models of military weapons, it is unusual to find any flint-lock provided with a swivel ring in the butt, as is customary with modern military arms. In early military models, the wood forend reaches almost to the muzzle and is secured by pins, but in later types it shows the influence of the duelling pistol design and reaches only half-way, and is bound to the barrel by metal rings.

Previous to 1780 the usual hammer was of the

swan-neck type, except in miniature pistols and some military models. The thin neck of the hammer was one of the weakest points in the weapon, and had of necessity to be made of fine quality steel. It was superseded by the double-necked or loop-hammer, in which the screw that tightens the jaws of the cock upon the flint passes into a double loop, giving it additional strength to resist the blow upon the frizzen.

In first-class duelling pistols the old swan-neck was long retained, but in early nineteenth-century flint-locks the loop-hammer, in a neater and more compact form, is more usual than the other.

The difference between duelling and horse pistols is often slight, many arms being made for use in either capacity, but bores of over twenty-eight—that is, half an inch—were seldom used for duelling purposes. Against this must be set the fact that pairs of pistols of duelling pattern were made for officers in seventeen bore, to take the Government ammunition of the period, and it is doubtful if these were not in frequent use for military duels of the period.

Brass or gun-metal barrels were popular for military and horse pistols, but never for duelling, and the last type of military officer's pistol abandoned the octagon type of barrel for a very heavy round barrel with a flat along the top for sighting, a type which was popular for house and carriage pistols.

Pocket pistols followed the trend of improvements in the larger pieces, and are usually of top hammer action, with enclosed lock, folding triggers, unscrewable barrels for loading at the breech, and a sliding safety bolt that locks the cover of the flash-pan to prevent the accidental escape of the priming while in the pocket. The same model, but with ordinary trigger and trigger-guard, was equally popular.

The pocket pistol was always the subject of experiments in the direction of greater rapidity of fire, and to it were applied a variety of queer attempts at double and repeating mechanisms.

The only one of these that seems to have had any vogue was the tap mechanism, in which barrels were arranged one above the other.

The top barrel was fired in the ordinary way; then the pan was closed again, the hammer re-cocked, and the tap turned. This tap served as a false bottom to the pan, and being turned through 180 degrees, exposed a second charge of priming lying in a recess in the tap itself, and communicating with the lower barrel. The pistol was fired a second time, thus discharging the lower barrel.

There is a piece by Knubley of London, in the author's possession, which has this principle, but four brass barrels arranged in under and over pairs. The right-hand pair are kept isolated from the left by a sliding-pan cover, and the functioning of the weapon is—Left top barrel fires; turn the tap, left lower barrel fires; pull back slide, right top barrel fires; turn tap back again, lower right barrel fires. The frizzen has to be closed, and the hammer recocked for each shot. A patent for this type of weapon was taken out by Simmonds of Birmingham, a notable maker of brass barrelled pistols in 1770.

A favourite attachment to pocket pistols is a spring bayonet lying on top of or along the barrel. When a catch is released, this springs out and locks in position, allowing the weapon to be used as a dagger. This device is the patent of John Waters, of Birmingham, in 1781.

Pocket pistols of a cheap type were made by wholesale firms and sold to gun shops all over the country. They are seldom of much value, unless of first quality and well decorated, and are not often found in cases.

The exact dating of pistols is a very complicated subject, owing to the practice of bringing the family weapons up to date. The old pistols were so good that it was frequently thought desirable to bring them up to date with a new lock, rather than abandon the beautiful workmanship of stock and barrel as obsolete. I have by me as I write an Irish pistol with a stock of the period of 1720, a long barrel of the same period proved and marked "Dublin," and a lock of some seventy years later by Lord and Hutchinson, engraved to match the original style of the piece. Conversions of flint-lock to percussion are common, particularly with expensive silver-mounted pieces, and such conversions detract considerably from the value of the piece to a collector.

Sometimes an idea of the date of manufacture can be gained from the hall-marks on the silver mounts or masks upon the butts, but the author has seen specimens of late percussion model pistols with silver mountings of a century before, so little reliance can be placed upon such dating.

Proof-marks are of value, but many pieces of

early manufacture lack these, and they give no indication of the date of manufacture, and it is only by experience and a knowledge of the difference of type and design, mountings, and manufacture, that one can approximately place the date of any definite specimen.

To aid such research I have compiled a list of makers with the periods at which they were in business. It is incomplete, but anyway a beginning. Owing to the fact that many retailers styled themselves "gunsmiths" and makers, it is not easy to tell if some of them actually made as well as sold pistols, but between the proof-marks the maker's initials or trade-mark often occurs, and these sometimes fail to tally with the name of the barrel.

Pistols are often found with different names on the lock and on the barrel. This is due to later alterations, and is not necessarily a defect.

Pistols with left-handed locks are uncommon, and any unusual variety of flint-lock pistol of European manufacture, be it multi-barrelled or unusual in design or finish, is, as a rule, well worth collecting, but to picture or describe unusual pieces is outside the scope of this little work. Single specimens of pistols that go in pairs, like duelling pistols, are not of so much value or interest as complete pairs with case; but a good single specimen is always better than a bad pair.

When pairs are in cases it is desirable to have all the accessories complete and correct. In nine cases out of ten the bullet mould found in the case is not the correct size, and often the powder-flask, is of a later date.

Now, for collectors' purposes, it should be seen to that all the fittings in the case, and the case itself, suit and are adaptable to the weapons in question—a perfect outfit complete in all details being the only correct way of keeping pieces.

I always make a point of purchasing any old bullet-moulds, powder-flasks, or fitments, that I come across in order to replace missing or incorrect pieces.

A complete case should contain a bullet-mould, a powder-flask, a cleaning rod, patches, bullets, turnscrew, spare flints, and pewter oil bottle.

In very complete sets a cutter for cutting out the patches, and a small priming flask, or miniature of the big flask, are also included. Spare flints can be got from the flint knappers at Brandon, or from most good curio shops where arms are sold.

Powder-flasks of the flint-lock period are always pear-shaped, made of thin copper, and with a spring cut off. To load, a charge was measured out by placing the finger on top of the spout, holding the flask upside down, and pressing the spring for an instant. The flask was then righted, the finger removed, and the charge poured down the barrel.

A bullet free from excrescences where the stem from the mould had been cut off was next selected, placed on a circular waxed cotton patch on top of the muzzle and forced down upon the charge with the ramrod.

The pan was next opened and a small charge of fine grained priming powder poured into the recess, the pan closed and the pistol turned lock uppermost, and tapped to allow the powder to run into the touch-hole.

It was then in a condition to fire.

With cases of percussion pistols, a variety of other accessories are also found. These are nipple keys, prickers (these are often combined with the turnscrew), and, in the case of muzzle-loading rifled pistols, a mallet and rammer to drive the bullet down. Late Continental duelling pistol sets have a variety of other small fittings as well.

The flask of the percussion period is not pear-shaped, but oval, and flat-bottomed, and is usually provided with two internal chambers to take a few bullets, and some caps as well. Many of the best of these were made by Sykes, and, though comparatively modern, are already getting scarce.

Bullet-moulds are provided with a small pair of cutter blades at the back of the mould. These are for the purpose of cutting off the projecting spigot of lead after casting.

The ends of the handles are often shaped as turnscrews, barrel wrenches, or ramrods, and most moulds have the size of the bullet they cast stamped on the handle near the cutter.

The flint-lock period is the richest for the collector, but it is only worth while collecting good stuff. The antique shops are filled with highly ornamented rubbish : genuine flint-locks, it is true, and highly decorated, but nine-tenths of it is Turkish and Greek trade "antiques." Flint-locks of the most valueless description were exported for the Eastern market, where they received crude decorative treatment with brass wire, base silver, and pewter, inlays, and ornament ;

they then return to Europe as "genuine antiques," and sell for about thirty shillings apiece. Few of these have ever been fired; none of them are meant to be, but the collector should beware of all Turkish, Albanian, and similar pieces; in nine cases out of ten they are pure fakes, built out of scrap for the antique market.

Good Oriental pieces are scarce, and a great deal of knowledge is required to distinguish between the good and the bad. The collector will be well advised to neglect this branch until such time as he has acquired a sound knowledge of European pieces.

CHAPTER V

THE PERCUSSION SYSTEM

THE revolution effected by the discovery of the percussion system was not an instantaneous development following the invention, but one which took quite forty years to become universally popular.

In 1807 the Reverend Alexander Forsyth, a Scotch clergyman of Bethvie, patented a system for utilizing fulminating salts to fire powder, and further specified as one of the chief advantages of his system that it prevented any escape of gas through the touch-hole. Forsyth's invention was not the usual copper cap of later date, or any metallic primer, but consisted of a device attached to the top of the pistol by which a few grains of fulminating powder fell by gravity into a cavity beneath a plunger and over the touch-hole. This plunger, struck by a peg in the nose of the hammer, ignited the charge.

This invention was marketed in 1812, but involving as it did the use of loose priming powder of an easily detonated variety, was neither safe nor popular.

The principle of ignition by percussion attracted universal attention, and gunsmiths in all parts of the world turned their attention to the matter. The result was a series of devices—detonating pellets, tubes, pastilles, and paper caps, as used in toy pistols to-day, or in connected tapes or strings for special primers. All were tried and abandoned in favour of the copper percussion cap, which many makers claim to have invented. Colonel Hawker claims it for Joseph Manton; Mr. Walker, of Philadelphia, has a good right to it! Egg also claimed it, as did many other makers of renown, but by 1815 the copper cap was in limited use on high-class private arms, and by 1830 it had ousted all other systems, and was recognized by both the gun trade and sportsmen as the best and only system of ignition for modern use.

Both flint and percussion weapons were in use at the Battle of Waterloo, but the newly invented percussion pieces were the private property of the officers who carried them. The rank and file still retained the flint-locks, and the very same weapons as were used at Waterloo were used in the Crimean Campaign, forty years later.

It is said that a young officer, William Spain, at that time attached to the Staff, had the ill-fortune to incur the Duke of Wellington's censure for not carrying a pair of pistols that would take the service ammunition. The pair he was carrying were his percussion duelling pieces made by Fiske of the Strand, and with the dates and occasions of his having been "out" engraved on the silver plates on the butts.

Contemporarily with the percussion system, breech-loading devices came into notice, but owing to escape of gas at the breech none of these were satisfactory till the cartridge had been invented. Numerous experimental weapons are, however, in existence, and pieces of this kind are of value to collectors.

The percussion-lock in its simplest form substitutes for the cumbrous flint-holding hammer and external flash-pan a simple steel nipple communicating with the powder chamber. Upon this nipple is placed a copper capsule or cap, containing a minute charge of fulminate of mercury, or other chemical compound, which detonates or explodes when hit. This cap charge is covered by a disc of tin foil and a drop of varnish, thus rendering the cap more or less waterproof. The fall of the hammer upon the cap crushes the detonating powder between the hammer and the nipple, causing it to explode and fire the charge. The cap held by the hammer over the vent-hole of the nipple prevents any escape of gas, as was inevitable with the open touch-hole of the flint-lock, and a better and speedier explosion, with no waste energy, results. In wet weather it was impossible to use a flint-lock, as the priming got damp, and in windy weather misfires were frequent, owing to the blowing away of the priming powder. The copper cap did away with all these disabilities, and increased the certainty of ignition, and decreased the risk of misfires to an incalculable extent.

The new system was universally approved, and by the simple process of fitting a new hammer and removing the flash-pan and cover to substitute a small butt-ended tube screwed at right angles into the enlarged touch-hole of the old barrel, and mounting a nipple upon the tube, flint-locks could be converted to percussion. Converted locks are easily recognized, even if complete new barrels have been fitted, because the hammer has a much longer fall than is usual with the lock built for percussion ignition.

It is seldom possible to reconvert such altered weapons, and never advisable, as modern restorations are never as good as the original work, besides spoiling the genuineness of the piece.

The application of the new system did not modify the established style of pistols, but as rifling—an invention dating from 1498—came into fashion with the development of the percussion system, the new weapons were more often rifled than not. All types of pistols adopted the new lock—horse pistols, duelling pistols, and pocket weapons—and the best of all were fitted with rifled barrels and hair-triggers, adjustable back-sights, and all possible mechanical refinements.

The hair-trigger mechanism is typical of this period of duelling weapons, though in later days its use was forbidden. The principle of the hair-trigger is practically that of making the trigger itself into a separate lock, with a very strong spring action on a flying arm of the trigger, which actuates the scear of the pistol lock. The hair-trigger was invented at Munich in 1543, and had long been applied to Continental rifles. Pistols with this action can be recognized by their having a small capstan-headed screw just in front of the trigger, and within the trigger-guard. This screw is the fine adjustment of the hair-trigger action.

The lock is so designed that it can be used or the pistol can be fired by ordinary trigger pull, according to the desire of the shot.

The action is set by pressing the trigger forward. This compresses the spring. A very light touch on the trigger releases its scear from engagement, and in place of the steady pressure from the human finger, it is driven back by the spring against the pistol lock release, the tail of the scear, and fires the pistol with a quicker and clearer action than is possible with the finger pull.

At this time, too, the saw-handle duelling pistol came into favour, but this type of handle, in which the shape was exactly like a tenon saw, was not popular for other uses. I have seen a small pair of all-metal saw-handle pieces with belt hangers and stirrup ramrods by Manton—not Joe Manton, but his successors; but these were probably a special order executed possibly for an officer of the East India Company, as the style was similar to weapons popular in the Indies.

The percussion system rendered pocket arms much less cumbrous and much cheaper to manufacture. Soon a quantity of percussion pocket pairs were on the market, varying from neat little folding trigger weapons, about four inches long, to eight-inch pieces of large bore. None of these had extensions of wood up the barrel, and they were nearly all reversions to the box-lock model, with top hammer (or in better classes "side hammers") ; that had always been popular.

The earlier percussion duelling pieces were like the best flint-locks, and had wood up to the muzzle; but soon the better class makers cut down the fore-end to half-way up the barrel, where it had a horn tip ; and some altered the style of the locks.

Till 1825 most locks were "side action"—that is to say, they had the main-spring in front of the hammer, and underneath the nipple. These gave place later to "back action" locks, which had the spring behind the hammer along the stock. By 1840 these were the most popular, as they did not weaken the woodwork of the pistol as much as the old system.

The very earliest percussion pieces had the lock secured to the stock by two long screws traversing the stock, exactly as was the practice with flint-locks. In very late flint-locks and most percussions, one screw (called the "side nail") alone suffices to hold the lock, while its forward end hooks in under an internal screw head recessed into the wood.

Nearly all first-class duelling pistol locks have a detent and a safety catch integral with the lock mechanism.

There were many and various patents applied to the percussion period. One of the best known of these was Nock's patent breech, in which a separate chamber was formed between nipple or touch-hole and powder-chamber. Here a few grains of powder found their way, and the flash of the cap had room to expand, insuring a better ignition. Nock's best pistols—and there was no better make at the time—have these curious and rather heavy breeches, and they are sometimes mistaken for conversions. Nock's patent breech was, in many ways, a reversion to the old cannon-barrel type, where a separate chamber for the powder was provided. The main point insured by this chamber was that the ball when rammed down did not compress the charge into a solid block that would delay the action of the flash. The same principle was used in mortars.

At this time queer pocket weapons and repeating and multi-shot devices of all kinds attracted the attention of inventors.

The large-bore small-sized pocket pistol known as the " Derringer " was first a muzzle-loading percussion pistol, and another type was the " Delvigne," with concealed hammer and stud-trigger. This latter piece was breech loading, but smooth bore, and similar to the modern automatic in appearance.

Pistols with safety devices for holding the cap on the nipple, weapons with concealed hammers, under and over barrels, and all sorts of queer oddments, made their appearance, but design had in the main settled down upon sound lines.

With the percussion duelling pistol the development of this type of weapon ceased, for the modern duelling pistol is to-day identical in all respects with those of sixty years ago.

The standard " Gastin-Renette " or " Bastien-Lepage " duelling pistols used in France to-day are smooth-bored muzzle-loading percussion weapons, and it is well to bear in mind that breech-loading or rifled pistols are not allowed to be used under the Duelling Code.

German pistols in pairs, with a bastard Remington action, are sometimes sold or advertised as " duelling pistols "; as, too, are often pairs of large-bore rifled or smooth percussions, which are properly horse pistols or house defence weapons. The collector should distinguish carefully between these and true duelling pieces.

The rifled duelling pistol has been barred as such

for many years; the round-bulleted smooth-bore being judged amply dangerous enough for the purpose of an affair of honour.

The mode of loading rifled percussion pistols was of necessity cumbrous; for after the powder charge had been rammed down, the bullet—spherical or conical—was enclosed in a circular greased linen patch and hammered down upon the powder against the resistance of the rifling by means of a ramrod and mallet. In modern duelling pistols sets in cases a mallet is supplied for this purpose; other fitments include screws for withdrawing the charge, powder-flask, bullet-mould, ramrod, nipple key and a pricker for cleaning blocked nipples. The best sets of the period have sometimes as many as three sets of barrels: one pair smooth bore of duelling length; one a long rifled pair for target work; and the third pair short for carrying in a carriage or upon the person.

The accuracy of good percussion pistols equals that of a rifle, and wonderful practice can be made with them.

The principal development of the percussion system as applied to pistols was the perfection of the revolver—a matter dealt with in the following chapter. At this point, muzzle-loading and breech-loading developments overlap; and as the connection is fully covered by the subject of "Revolvers," it will be better to take these first, and single-shot, breech-loading pistols later on, despite the fact that muzzle-loading revolvers belong of rights to the chapter on "The Percussion System."

CHAPTER VI

THE DEVELOPMENT OF THE REVOLVER

From the earliest times inventors have utilized the revolving barrel or chamber principle in order to effect a repeating firearm, but in nearly all cases the rotation of the fresh chamber or barrel was effected by hand rather than by the action of trigger or hammer.

The first type of revolver, apart from museum specimens of great rarity, such as the flint-lock snaphaunce revolver in the United Service Museum, is the familiar "pepperbox," a design of four or six barrels rotating around a central spindle, and fired by the blow of a hammer actuated by the trigger which also revolved the barrels in exactly the same manner as the modern double-action revolver. There were many makers of these, such as Allen and Thurber, Walker's patent, and many English and Continental firms, and several types, some with hammers above and some with them below the barrels. Some models have separate barrels, but the majority are bored out of one solid piece. A notable variation is the "Comblain," in which the barrels remain stationary while a revolving internal hammer strikes the caps.

In all these "pepperbox" revolvers the nipples

were at right angles to the cylinder, and it was not until after Colonel Colt in 1835 patented his first revolver that centrally placed nipples were employed.

For a complete account of Samuel Colt and his extraordinarily successful career as an inventor and as a manufacturer, the reader is advised to turn to Sawyers' excellent book, vol. ii., of "Firearms in American History," which gives complete particulars, and traces the evolution of the Colt revolver through its chequered career, from the first model "Texas pistol" of 1835, through the Mexican War model of 1847 and the great lawsuits over patent rights, then through all subsequent models of revolvers up to 1865. The book is voluminous in detail, and all models are illustrated, as are many other types of American historical arms and revolvers.

The chief points about Colt's patents were—First, rotating the cylinder by the act of cocking the hammer, so as to bring successive chambers in line with the barrel; second, locking the cylinder in the proper position during the moment of discharge, and unlocking it by lifting the hammer when cocking; third, that each nipple and its caps were isolated from the others by partitions on the cylinder, to prevent simultaneous discharge of two chambers by the flash of the cap. The London factory for Colt's arms was opened in 1853, but only worked till 1855, when it was taken over by "The London Pistol Company," and produced their replicas of "Colt" revolvers with the original machinery. The principal output of the London factory was the

1851 model Navy Pistol Colt revolver and the 1849 Pocket Pistol ·31 and ·34.

The number of Colt muzzle-loading revolvers of various types is legion, and they are mainly of interest to American collectors. Small details of size of trigger-guard, method of fastening the barrel, place of manufacture, etc., make considerable but fluctuating variations in their value, but Colt revolvers with secret triggers similar to the specimen in the United Service Museum, and models made at Paterson, N.J., are of considerable value.

The original and usual models are of ·44, ·36, or ·31 calibre, and have cylinders engraved with "stage-coach hold-up" or naval scenes. They are single action, and have the barrel simply connected to the central spindle around which the cylinder rotates, without any top strap attaching it to the breech-block. Loading is effected by ramming a conical bullet with a varnished gold-beaters' skin cartridge full of powder attached to its base into each chamber. This ramming, as is common with all muzzle-loading revolvers, is effected by means of a lever and plunger attached to the side of the barrel. Tight ramming was a necessity, inasmuch as the explosion of one chamber might fire another charge in the adjoining chamber if the loading had been with a loose ill-fitting bullet. The loaded chambers were then capped at the base, the cylinder being free to revolve when the hammer was at half-cock. Sighting was accomplished by means of a notch cut in the hammer, which, when raised, just intersected the line of sight. A contemporary opinion of the value of the new invention can be gleaned from Hans Bush's "Rifleman's Manual," and there is no doubt but that he correctly represents the feeling of sportsmen of the period. He states :

"I may now briefly advert to the 'repeating principle,' as applied to firearms in general, but more especially to pistols and rifled carbines. It is to our Transatlantic friends that we are indebted for the perfection of these weapons, for though, more than two centuries ago, various attempts were made to produce a series of successive discharges from one arm without the necessity of reloading, it is to Colonel Colt's perseverance, energy, and mechanical skill that the merit is due of having successfully vanquished all the difficulties that presented themselves in their construction."

Innumerable were the objections he had to contend with at the outset. Many sneered at the idea as preposterous. "They would always be liable to get out of order." "They would take too long to reload." "They would, besides, always be missing fire," etc.

As regards the liability of the revolving pistol to get out of order, this was satisfactorily disproved by a severe trial instituted by order of the Board of Ordnance of the United States, who directed a holster pistol to be discharged *twelve hundred* and a belt pistol *fifteen hundred times*, cleaning them but once a day ; after which ordeal neither of the pistols appeared to be in the slightest degree injured.

With respect to the cost of production, as

almost every part is formed by machinery, hand-labour being only required in the finishing department, Colonel Colt seems likely permanently to retain in his own hands the business which his ingenuity has created, for he will, of course, always be in a position to undersell any imitators that may appear. Great security is also obtained from the cause above stated, for we find that upon "proof" only one barrel and one cylinder burst out of 2,082. The most perfect uniformity of detail is attained from the mechanism employed for the several parts of each class of weapon being precisely similar; so that if any become damaged on service, a great number of available arms can be immediately compounded of those which have been partially injured.

The ramrod attached to these pistols consists of a very clever but simple compound lever, which forcing the ball effectually home, hermetically seals the chamber containing the powder, and by the application of a small quantity of wax to the nipple before capping, the pistol may be immersed for hours in water without the chance of a miss-fire.

The movements of the revolving chamber and hammer are ingeniously arranged and combined. The breech, containing six cylindrical cells for holding the powder and ball, moves one-sixth of a revolution at a time; it can only be fired when the chamber and the barrel are in direct line. The base of the cylindrical breech being cut externally into a circular ratchet of six teeth (the lever which moves the ratchet being attached to the hammer) as the hammer is raised in the act of cocking, the cylinder is made to revolve, and to revolve in one direction only. While the hammer is falling, the chamber is firmly held in its position by a lever fitted for the purpose; when the hammer is raised, the lever is removed, and the chamber released.

So long as the hammer remains at half-cock the chamber is free, and can be loaded at pleasure. The rapidity with which these arms can be loaded is one of their great recommendations, the powder being merely poured into each receptacle in succession and the balls then dropped in upon it, without any wadding, and driven home by the ramrod, which, of course, is never required to enter the barrel.

While carried in the pocket or belt there is no possibility of an accidental discharge of these pistols. Whenever it is required to clean the barrel and chamber, they can be taken to pieces in a moment, wiped out, oiled and replaced.

The hammer at full-cock forms the sight by which aim is taken. The pistol is readily cocked by the thumb of the right hand, a plan in every way far superior to the arrangement whereby the hammer is raised by a pull on the trigger. This is in every respect most objectionable, the pull materially interfering with the correctness of aim, and the scear-spring having the duty of the main-spring to perform as well, is apt constantly to be getting out of order. *Not so Colonel Colt's; as regards the purposes for which they are intended, they may be pronounced in every respect perfect.*

The following testimony of Major Robert Bruce in favour of Colt's revolvers is quoted from the report (1854) of the Select Commtitee on Small Arms. It proves sufficiently their value and efficacy. "I consider," said that gallant officer, "Colt's revolver very far superior to any other we saw at the Cape during the Kaffir War. That opinion is founded upon its accuracy, and in its not failing in what you require. I have never seen an accident with it, and I have never seen it, in discharging, become involved in any difficulty. I have seen it fall into the water and returned to its belt without being discharged, and afterwards, perhaps after the lapse of some days, I have seen it discharged most perfectly without the least failure of any one of the barrels. I have ridden hundreds of miles with a Colt's revolver by my side, and always found that there was an impossibility of the powder shaking out. The lever ramrod forces down the ball in such a way that no other ramrod could do." Much additional evidence was given to the same effect—in short, no valid objection can be raised against them.

His cavalry pistols are, in fact, pocket rifles. With one of them I once fired from a rest thirty-six rounds at the enormous range of *four hundred and ten yards!* Six bullets struck the butt at distances varying from thirty to thirty-six inches from the centre of the target, eighteen bullets struck within the circumference of a circle seven feet in diameter, and the other six shots at heights varying from ten to twelve feet above the target — satisfactorily proving the capacity of the weapon for still greater range.

It must be remembered that for this experiment I had no sights adjusted. The elevation was taken altogether by guess. I am convinced, however, that if requisite sights could be fitted one might insure fair shooting, even with these seven and a half inch barrels, at 500 yards.

The wound which a conical bullet from one of Colt's revolvers inflicts is terrific, driving before it, as it does, a cylindrical plug of muscle or bone. The hæmorrhage or shock to the system is so great that death in the majority of cases usually ensues. No one, indeed, who has not had some experience with these diminutive rifles can form any idea of their capabilities. But their reputation is now so thoroughly established that the inventor may feel secure against the many attempts that have been made to impugn their character.

The success of Colonel Colt's revolver stimulated numerous gunmakers and inventors to further efforts, among which were one or two makes, notably Cockran's and Wilkinson's, with chambers arranged like the spokes of a wheel, the cylinder being vertically pivoted. In the Wilkinson revolver the cylinder was removable for independent loading by lifting a top strap that held it in position, and the firing was effected by a long hammer similar to the pepperbox pattern overreaching the rear chambers. Similar pistols, but with the radial cylinder vertically pivoted, have also been manufactured; but the system was never popular or successful, probably because of the mechanical difficulties and the distinct objection of the ordinary man to employing a weapon in which, at least, *one* loaded chamber pointed directly at him. By 1853 design had steadied down to the solid-frame muzzle-loading revolver of Deane and Adams or Tranter pattern, and large numbers of these weapons were despatched to Australia and the various British Colonies then in process of exploitation. In America an enormous production of weapons was taking place, and the revolver has always been recognized as their national weapon. At the same time, inventors were attempting to embody the cartridge and cap in one, in order to perfect breech-loading mechanisms, and the result of this is an overlapping period of percussion arms of the plain muzzle-loading type. Weapons made breech-loading to take a separate cartridge, but fired by means of a cap placed upon the external nipple, self-igniting cartridges of the needle-gun type, and, finally, the development of the metallic cartridge in its first varieties as rim and pin-fire systems.

Among the notable muzzle-loading American pistols we find the Savage, a revolver operated by pulling back an under lever to cock the hammer and rotate the cylinder, a trigger connected with the lever being fired by pressure of the first finger, the lever itself being operated by pressure of the second. In this weapon the cylinder was held up against the base of the barrel by a method similar to the later invented Nagant.

The Starr revolver, which had a tip-up barrel secured by a pin fastening in the top of the breech-block, and which could be "broken" in order to remove the cylinder for loading or cleaning.

Allen and Wheelock revolvers with a side-action hammer striking tangentially set nipples, and a host of other makes, such as Remington's, Whitney's, Roger and Spencer's, Springfield, etc., which in the main followed Colt's design with the exception that most of them were solid framed—a top strap connecting the butt of the barrel with the top of the standing breech.

Metallic cartridges of the rim or pin-fire types made their first appearance about 1850, for at the Great Exhibition of 1851 we find the French gunmaker, Lefaucheux, who invented the pin-fire cartridge; Flobert, the inventor of the bulleted breech-cap and the father of the rim-fire system; Lancaster, with his central-fire cartridge, used in walking-stick guns to the present day, and the Continental needle-gun cartridges.

These needle-gun cartridges were the first type of self-igniting cartridge used in pistols, and consisted of a bullet having a cap or detonating pellet in its base; behind this was the powder charge, in a combustible linen or paper case, the base of which contained a wad pierced by a central hole and lightly covered with blue waterproofed paper. Ignition was effected by a hammer carrying a long pivoted needle, which struck through the base wad and powder charge, igniting the cap at the base of the bullet. The base wad was pushed forward by the next cartridge on entering the chamber.

A few Continental pistols for this type of cartridge, with a tube container above the barrel and a crude falling-block mechanism, are still extant, but the ammunition was dangerous, and the escape of gas at the breech alarmingly unsafe for a repeating mechanism.

The pin-fire cartridge is substantially the same as is used to this day in cheap Continental weapons, but was more dangerous and less reliable than the rim or central fire; it also suffered from the disadvantage of having to be placed in the chamber with the pin in its proper position projecting from the notch, and as this position in a revolver is of necessity vertical, the hammer and cocking-piece interfered with the sighting of the weapon.

The rim-fire cartridge originated from the development of the saloon or small-bore target pistol, and the author has been fortunate enough to collect at various periods, pieces showing the development of the idea.

The first step is the muzzle-loading ·22 calibre weapon with a nippled breech, the action being opened to admit of the placing of a percussion-cap upon the nipple set in direct line with the axis of the bore. The second step is the substitution of a hinged anvil and extractor, thus making the piece entirely breech loading; while the third step is the plain, rimmed-bulleted breech-cap. This idea was readily seized upon by American inventors, and the evolution of the early Smith and Wesson revolvers of ·22 calibre, the barrel hinging up from beneath, the Colt, Ballard, and Remington "Derringers," the Sharps four-barrelled pistol, and kindred rim-fire pistols of American manufacture, was the direct outcome of the invention of bulleted breech-caps.

By 1860 breech-loading pocket weapons were popular, and small-calibre repeating mechanisms, such as the Remington repeating Derringer of ·32 calibre, the Henry Volcanic pistol, which was the forerunner of the Winchester rifle, and a number of similar inventions, were in use. Breech-loading for the larger bores was not common at the period of the American Civil War, but came into fashion very shortly after.

The development of the Derringer from the large-bore conically rifled pocket muzzle-loader into the ·41 rim-fire, single-shot, vest-pocket, self-defence pistol is peculiarly interesting. The first Colt Derringer has the same pivoted swing-over mechanism as was common in the first French saloon pistols of the same period. This first Colt Derringer (No. 1) was made in metal throughout, and pairs were issued for the use of Indian Civil servants. The weapon has an almost concealed hammer and a "knife blade" extractor to loosen the fired cartridge. The second Colt Derringer (No. 2) was similar, but had wooden stocks and a rather larger grip, while the third is the present-day model, with spring ejector and horizontally pivoted barrel. Among other American Derringer pistols were the Ballard, a tip-up weapon with a ratchet extractor similar to the one in early model Smith and Wesson Frontier revolvers and the double-barrelled Remington Derringer as used to-day. This weapon has an automatically movable striker on the hammer, which moves when the hammer is cocked, and strikes the rim of the cartridge in the upper and lower barrels alternately.

Derringer pistols were not built for accuracy at long ranges, but for close work over a table or shooting through the pocket. For this purpose of self-defence the heavy bore (·41) and powerful cartridge were eminently suitable. All these pistols and most American revolvers of the period were single action, and with the majority no trigger-guard was fitted, popular prejudice being distinctly against any unnecessary fitment that might delay one in an instant drawing of weapons to exchange shots.

The Sharps four-barrelled pistol was a popular model at this period, and was mostly made in ·22 and ·32 rim-fire calibres. A revolving nose upon the hammer caused the striker to fire the barrels in rotation, a device similar to that employed in the Comblain and Braendlin four-barrelled mitrailleuse pistols of later date.

The adaption of muzzle-loading revolvers to breech-loading systems brought out some queer devices, and in some cases merely the nipple breech of the cylinders was turned off, leaving the ratchet as a spindle projection. Around this boss was adapted a keyed ring fitted with strikers, which were hit by the hammer and fired the central-fire cartridges in the ordinary way without further alteration to the hammer or standing breech. Later models fitted a larger cylinder and a striker to the hammer. Pistols converted in this way are still in use in Mexico.

The Colt ·45 Frontier six-shooter, of enormous fame, was the first and most celebrated of all central-fire breech-loading revolvers, and maintains its popularity as a reliable, accurate weapon to the present day.

The alteration from the muzzle-loading revolver was slight. The barrel and standing breech were connected by a top strap, or rather the barrel was screwed into a solid frame; the hammer was modified and equipped with a striker, and the loading lever gave place to an ejecting rod. Loading was accomplished through a gate at the right-hand side of the frame, and a notch upon the standing breech replaced the hammer backsight. The grip and balance of the weapon remained almost unaltered, though the single-action Bisley model ·45 calibre Colt, with long handle and small hammer, was a distinct advance upon the original model.

Smith and Wesson, who had started to make breech-loading revolvers in 1859, next produced their celebrated self-ejecting models, the first patterns of which had ratchet ejectors and gracefully curved handles similar to the Colt grip. In later models this grip changed to the well-known Smith and Wesson shouldered handle, and the ejector assumed the present cam form similar to the Webley system.

These early weapons were all single action, but a demand arose for "self-cockers," or double-action revolvers, and the Colt Frontier double-action and "Bull Dog" models met the demand. The Frontier types of pistol were usually chambered for the ·45 Colt or the standard Winchester rifle cartridges of ·44-·40, ·38-·40, ·32-·20 calibres. This was to enable the frontiersman, cow-puncher, or miner, to use the same standard cartridge in either rifle or revolver. The "Bull Dog" models were chambered for the ·41 long and short Colt's central-fire, the ·38 and ·32 Colt cartridges, but single-action pocket revolvers, with solid frames and a pretty little seven-shot ·22 calibre pistol, were also made to take rim-fire cartridges.

Other makers of this period were Smith and

Wesson, Marlin, Merwin and Hulbert, Remingtons, Hopkins and Allen, and other arms companies too numerous to mention.

In England the solid-frame R.I.C. model Webley's, Adams', and Tranter's, of similar pattern, developed on similar lines, the double-action revolver having always been more popular in this country than in the States.

In Europe the various countries evolved special military weapons, such as the French and German Ordnance revolvers, but it was not until 1890 that the self-extracting Mark I. Webley and its variations in smaller calibres were officially adopted.

Of all the pistols up to 1890 the Colt single-action Frontier was the most popular for general knock-about use. The Bisley model and the self-extracting Smith and Wesson were rivals for target shooting, while the English double-action military and constabulary revolvers were much more in favour than the American double-action models.

This period brings us to the last twenty years of pistol production, and closes the historical section of this book. The standard design of the modern revolver begins with the self-ejecting Webley models, continued into the modern solid-frame, swing-out types of Colt and Smith and Wesson, and eventually gives place to the automatic pistols now made by all three firms and many others.

CHAPTER VII

HISTORY, WARS, AND THE PISTOL

"MRS. PAGE. Alas, three of Master Ford's brothers watch the door with *pistols*, that none may issue forth."

The Merry Wives of Windsor.

THE English Chronicles tell us that when Edward VI. landed at Ravenspur in Yorkshire, 1471, he brought three hundred Flemings armed with "hand gonnes," but it is probable that these were hand cannon, or guns, rather than anything of a pistol type.

Henry VIII.'s reign sees the first introduction of the wheel-lock dag, or pistol, somewhere about 1530, as an official English weapon, but it is more than probable that match-lock pistols accompanied the discoverers of America—Columbus in 1492, Cabot in 1497—and were in use during the European wars of 1500-1557.

At this period of history the Protestant Reformation and the divided state of the rich and artistically advanced Italy produced a continual state of European warfare. Francis I. of France was crushed by the Duke of Bourbon and Charles V. of Spain at the Battle of Pavia, 1525, contemporary pictures of which show powder-flasks and match-locks; and the succeeding period up to 1560 was one of continual warfare, during which firearms slowly replaced the pike and halberd, until a sixteenth-century army mustered a large proportion of musketeers, arquebusiers, and cavalry—the latter armed with primitive pistols. The mercenary German reiters used the pistol as a cavalry weapon, and spread its use throughout Europe.

In the reign of Elizabeth (1558 to 1603), Drake, Hawkins, and Raleigh waged war with Spain, and the pistol or petronel, as a convenient weapon for shipboard use, was already popular at the period of the Armada, 1588; while in 1584 William of Orange—William the Silent—was assassinated by means of a pistol shot fired by Balthazar Gerard.

Elizabeth encouraged the London gunmakers, and during her reign the number of gunsmiths increased largely, but during that of her successor, James I. (1603-1625), the trade fell into disuse till there were only five makers left in London.

The Gunpowder Plot conspirators, in 1605, fled after the failure of their plot to Holbeach in Lincolnshire, where they used pistols in their attempt to resist capture.

In France the wars of Henry of Navarre and the Holy League ravaged the country. The French cavalry, composed of troops of gentlemen, had adopted pistols at the time of the Battle of Ivry in 1590. At the Battle of Longside, in 1568, we get from a contemporary writer (Sir James Melville) a description that shows that the pistol was still used as a missile. "The long pikes" of Queen Mary's men "were so thick fixed in the others' jacks that the pistols and great staves thrown by them which were behind could be seen lying on the crossed weapons." The final combination of Henry III. of France and Henry of Navarre against the League, and the assassination of the former and the crowning of the latter Henry IV. in 1598 put an end to the civil war, during which the pistol was a recognized nobleman's weapon and a favourite arm for assassins.

In 1618 the Thirty Years' War broke out in Germany, and Gustavus Adolphus of Sweden, Wallenstein, and Tilly became the military leaders of the age. Gustavus Adolphus recognized the value of firearms and their great mobility, and under him the cavalry—supported by field artillery—were regularly equipped with a pair of pistols, a procedure that was followed by all European military commanders. His death at the Battle of Lützen in 1632 was attributed to a pistol bullet.

A "Military Treatise" written in 1619 gives some idea of the common firearm of the period:

"Hee that lovyth the saftie of his owne person, and delights in the goodnesse and bewtie of a peece, let him alway make choice of one that is double breeched, and if it be possible, a myllan peece, for they be of tough and perfecte temper, light, square, and bigge of breech, and very strong where the powder doth lie, and where the violent force of the fire doth consist, and notwithstanding trimne at the ende. Our English peeces approache very nighe unto them in goodnesse and bewty, (theyr heavinesse only excepted) so that they be made on purpose,"—*i.e.*, specially ordered—"and not one of those common sale peeces with round barrels, whereunto a beaten soldier will have great respect, and rather choose to pay double money for a good peece, than to spare his purse and endanger himselfe."

When Charles I. of England ascended the throne in 1625, he found the gun trade in a bad state, and forthwith issued orders in 1632 that a stated scale of charges for weapons should be drawn up. Under this we find that—

For a pair of fire-lock pistols *furnished with a key* (*i.e.*, wheel-locks), mould, powder-flask, worm, scourer, in cases of leather and of length and bore according to the allowance according to the Council of War	£iii	o	o
For a pair of horseman's pistols furnished with snaphaunces and similar equipment to the above	£ii	o	o

The Civil War of 1642-1648 was remarkable for cavalry work, and the pistols in use were principally wheel-locks and a sprinkling of snaphaunces. Nye, the gunner of Worcester (1649-60), and Henry Radoe, of Norwich, were both acquainted with the snaphaunce; the latter as early as 1585. In 1643 John Hampden was mortally wounded by a pistol bullet at Chalgrove Field during a cavalry skirmish, and Cromwell's regiment of Ironsides were armed with pistols. In Scotland the all-steel Scotch dag was in favour, and was very popular as an officer's pistol with the Cavaliers. The Scotch all-metal Highland pistol was a distinct type of flint-lock; yet its development undoubtedly bears traces of the influence of the excellent French pistolets of the period.

A very complete monograph and dated list of the best makers and marks is in existence, and a copy (in French) can be seen at the library of the Victoria and Albert Museum.

The village of Doune was a notable centre for armourers and a place of frequent cattle fairs, where the clansmen could exchange their cattle for goods.

The art of making these special pistols was introduced there about 1646 by Thomas Caddell, who left the trade to his children and apprentices.

A regular guild of pistol-makers sprung up there, and a large export trade was initiated, the Doune pistols being particularly appreciated in France and Germany, where half-armour still continued in use, the all-steel finish of the pieces blending well with the armour.

Among many of the best armourers of Doune were Caddell, James Sutherland, Michie, S. Micine, John Campbell, S. Stuart, Thomas Murdoch, John Murdoch, and David McKenzie. The trade did not die out till the beginning of the nineteenth century, and continued into the period of percussion-locks.

In 1798 Thomas Murdoch was the leader of the trade, and his prices varied from four to twenty-four guineas. The earlier types of Scotch pistol can be distinguished by their possessing a half-cock plate projecting from the lock in a manner similar to the release of a Spanish or Oriental lock.

The full-cock is, however, similar to the ordinary flint-lock, and cut on the tumbler inside the lock-plate. A very large bore wheel-lock dag for throwing hand grenades was used by some troops at the siege of Oxford.

After the Civil War many Cavaliers and Puritans emigrated to the newly founded colonies in America and the West Indies, and the pistol was a recognized naval side-arm during the Dutch Wars of 1652.

The restoration of Charles II. in 1660 saw the wheel-lock superseded by the true flint-lock, with which all of General Monck's battalions were armed, although on the Continent wheel-locks were still made in Germany and Italy.

James II. acceded in 1685 and was deposed in 1688. During the Monmouth rebellion in 1685 the rebels had a mixture of weapons, but the royal troops were well armed, the long flint-lock pistol of sixteen inches having superseded the long, two-foot wheel-lock or snaphaunce as an officer's weapon.

On the Continent flint-lock and snaphaunce pistols were adopted earlier, and at the siege of La Rochelle in 1628, under Louis XIII. and Cardinal Richelieu, the snaphaunce was in use.

By 1642 Louis XIV. and Cardinal Mazarin were the effective combination that steered France to a position of pre-eminence among the nations, and the style and decoration of French weapons of the Louis XIV. period is excellent, so good, indeed, that the same art was often applied to weapons of much later date and to English-made pistols by foreign artificers.

The accession of William of Orange (William III. of England) in 1689 was the prelude to a reign of warfare and rebellion. Scotland and Ireland were aflame with Jacobite sympathy, and the French under Louis XIV. assisted them with arms. In 1701 the Grand Alliance of England, Holland, Germany, Portugal, and Savoy was formed to prevent the union of France and Spain, and to preserve the balance of power. During this time flint and match-lock muskets were in use, but the design of the pistol progressed. The ensuing war of the Spanish Succession broke out under Anne, who acceded to the throne in 1702. By this time short-barrelled pocket pistols to fit the great "salt-box" pockets of the William and Mary and Anne costumes were common, and the large-sized, cannon-barrel, side-lock pistol was a favourite military officer's arm.

The war with the Battles of Blenheim, Ramillies, Oudenarde, and Malplaquet lasted from 1704 to 1709, and although military model pistols had settled down to the standard "Tower-lock Queen Anne" weapon, other patterns of Continental make were in use in the army, while the navy had a very mixed equipment of all kinds of pistols of varying bores.

George I. (1714-1727) had merely the Jacobite rising of 1715 and the siege of Gibraltar by the Spaniards in 1727, and the death of Louis XIV. in 1715 removed a menace to the peace of Europe.

George II. (1727-1760) had the rising of 1745; the War of the Austrian Succession, 1740-1745; and beginning of the Seven Years' War in 1756; and the campaigns against the French in India (1756) and Canada (1756) with Wolfe's capture of Quebec in 1759. The pistols of this period do not vary much, the same range of military, pocket, and holster pistols being popular; but as the century advanced the manufacture became more solid, the military affairs calling for good, reliable weapons. The American Colonies were largely supplied by English makers, but in 1740-50 commenced the manufacture of firearms on their own.

The reign of George III. (1760-1820) saw the culminating splendour of the flint-lock period, and the beginning of the percussion era, and it was during this reign that English gunsmiths attained world-wide celebrity for the excellence of their weapons.

The distinguishing factor of this reign was the loss of the American Colonies and their War of Independence, 1775-1783. The first shot in this war was the one that provoked the skirmish at Lexington, and it was fired by Major Pitcairn, from a Scotch all-metal pistol.

The Georgian military pistols are clumsy weapons of about fifteen inches long, locks stamped with "Tower" crown, G.R. and Broad Arrow. Their barrels are merely pinned to the stocks, while those French and Continental pistols of similar type are usually surrounded by brass or iron bands.

The French and Spanish allied themselves with the Americans, and naval conflicts in the West Indian waters and the sieges of Brest and Gibraltar further developed the trade in firearms, both military models and improved private designs being modelled upon the popular duelling pistol.

The French Revolution in 1789, with the execution of Louis XVI. and Marie Antoinette in 1793, together with the early Napoleonic Wars beginning 1796, and marked by Nelson's victory of Trafalgar in 1805 and terminated at Waterloo in 1815, brought a great many French weapons to this country, while the campaigns in the Peninsula (1808-1813) under Wellington, account for many Spanish specimens. The notable pistol of this era was the British duelling pistol and the holster and military officers' weapons made upon the same pattern. It is curious that the Duke of Wellington never possessed a pair of duelling pistols, and when called out by Lord Winchilsea in 1829 had to borrow a case from Doctor Hume, his attendant.

George IV. (1820-1830) and William IV. (1830-1837) usher in a period of development of the percussion system, and the birth of the revolver, while Queen Victoria (1837-1901) oversees the whole period of breech loading and the invention of smokeless powder. Automatic pistols belong to the last year of her reign, though their first real development is confined to that of her successor, Edward VII. (1901-1910), and their adoption into the Navy as a regulation pistol to the reign of George V.

Duelling continued till 1844 in England, and is still legal, though discountenanced, in France and Germany.

With the abolition of duelling, English makers ceased to manufacture the noted duelling pistol, and by 1865 the breech-loading revolver had taken place of all single-shot weapons, except a few target and saloon pistols made principally by Tranter of Birmingham.

The preceding précis of dates and battles, when compared with the types of pistol shown in the various plates, helps to fix one's idea of their period and the slow advance of the importance of the pistol as a military or naval weapon. Up to 1775, the weapon was not accurate even at short ranges, but the later duelling and similar models were sure of a two-inch bull at twenty paces. The usual duelling distance was ten, and military pistols were really for use at point-blank range. The development of the revolver as a war weapon was due to the American-Mexican War of 1846, and to a certain extent to their war against the Seminole Indians in Florida in 1835, where the repeating system struck fear into the hearts of the Indians, used only to single-loaders. The celebrated Texas Rangers, who kept the Mexicans in order and freed Texas from Mexican rule in 1836, were all equipped with revolvers, and the name "Tejano" is a synonym in Mexico to this day for any good revolver shot. In the Mexican War later on, in 1846, all officers and many men carried Colts, and at close-range fighting, such as the Mexican then preferred, the weapon proved marvellously efficient. The discovery of gold in Western America and Australia in 1848 and 1849 spread the private use of the revolver, and the American Civil War in 1861-65 saw both sides revolver-armed. Some troops, notably Mosby's Guerillas, were armed only with revolvers, two on the belt and two in the saddle holsters, and proved extraordinarily fatal in cavalry conflicts.

The officers bound to the Crimea in 1854 were all revolver-armed, though the cavalry pistol was the same old flint-lock of Georgian times. The Indian Mutiny (1857) was well in the percussion period, though by 1867 (the Abyssinian Campaign) and 1882 (the bombardment of Alexandria) breech-loading revolvers were fully established.

In Europe no serious attempt has ever been made to teach the use of the revolver as a cavalry weapon, except among the Austrian cavalry who were armed with the Werndl pistol. It is still, as in the days of the Crimea, purely an officer's arm, or issued to small details such as despatch riders on special work. The reason for this is probably the prejudice with which the theory-trained European and British officer regards a revolver. Beyond firing the compulsory minimum course of so many rounds per year, they take, as a whole, no keen interest in pistols, and the advantages of the weapon for close work are completely ignored. Possibly they believe that the days of shock tactics for cavalry are over, and that horses are soon bound to be obsolete for European warfare. Even so, and admitting the necessity of mounted infantry work on the part of cavalry, in addition to their usual use, the revolver as a weapon in a skirmish is far better than sword or lance, and as a side-arm in addition to the bayonet for cyclist battalions, the need is obvious, for your military cyclist is at present unarmed when mounted, and merely equipped as an infantry-man with rifle and bayonet attached; while on his machine he must dismount, and is forced either to take cover and fight as an infantry-man or fly defenceless with the lance or sabre of his pursuer at his back.

The revolver would give him at least a sporting chance in hedgerow fighting, and the present cyclist battalions might well be armed with the

Webley's revolvers withdrawn from the Navy on the issue of the present automatic pistol—certainly it would cost the nation no more, and would be better than merely storing them or breaking them up.

The United States of America Army and Navy both have pistols as the official side-arm, and in Cuba and the Philippines the revolver proved its excellence as a military weapon.

Europe, in her blindness, is as ever well behind the times.

Author's Note.—Since the preceding paragraphs were written, the European War has brought the revolver once more into prominence, and the military authorities are again learning its value. I am glad to say that a better course of training is now insisted upon, and, though much still remains to be done, there are signs that the authorities are waking up.

CHAPTER VIII

AMMUNITION

THE development of sound ammunition has to a large extent been the reason for the production of modern pistols. The one has reacted upon the other, and where in some cases weapons were designed to fit special cartridges, in modern practice special cartridges are designed for improved pistols.

The modern ammunition maker's catalogue is a mine of detailed information about powder charges, bullet weights, initial muzzle velocities with different barrel lengths, average muzzle energies, and standards of penetration; and little more than a brief survey of popular pistol cartridges and their historical development can be given in these pages.

For more detailed knowledge the reader is recommended to refer to the catalogue of Kynochs, Eleys, The Winchester Fire Arms Co., The Union Metallic Cartridge Co., or any first-class ammunition firm. In cases where very special information is required the expert departments of these firms are usually exceedingly courteous in replying to inquiries, but it is hard to find any point not already covered in their annual catalogues. Modern cartridges are evolved by a process of calculation and experiment in which the weight of the weapon, the calibre, the maximum weight of bullet, and the length of barrel, are all taken into consideration.

The best powder charge having been found, different scales of powder charge adapted to various bullets, such as special flat-nosed target types and special hollow-nosed hunting bullets, require similar variations from the standard load.

Nowadays all modern pistol and revolver ammunition is made for smokeless powder, and a specially hardened or nickel-based or nickel-covered bullet is used, but the practice of giving the powder charge in terms of "grains of black powder" is still in use, as black powder forms a good normal standard of comparison.

In actual practice a charge of smokeless "equivalent to so many grains of black" is used, though in some automatic and modern military charges the load is given as "grains of cordite or dense smokeless powder."

To revert to the history of the cartridge, one must remember the development of the device, on the principle of embodying the percussion-cap with the charge to facilitate effective breech loading.

The muzzle-loading cartridges were either a spherical or conical ball, behind which was tied a small conical goldbeater's-skin bag containing the powder charge. This skin was varnished to exclude damp, and the whole was wrapped in a tough cartridge paper case from which projected a tape. Pulling this tape ripped open the paper envelope and exposed the fragile charge, which was rammed down the barrel or chamber, and in some cases further wadded home by ramming down the paper cartridge case on top. The nipple was then capped, or, in the case of flint-locks, the flash-pan "primed" with a small charge of fine powder.

Most muzzle-loading pistols were not used with cartridges but simply loaded with a charge of powder out of a self-measuring powder-flask, wadded down with a paper or linen patch or wad followed by a bullet wrapped in a patch, or if tight fitting forced down by means of a mallet In some cases a top wad was used to prevent the bullet rolling loose out of the barrel or not resting tight upon the powder charge.

Muzzle-loading revolvers were used with factory-made ammunition of skin cartridges, although sent out direct from the makers, complete with bullet-mould.

Some English makers issued revolvers for use either with the ordinary loose load or "Daws Copper Cartridge," which consisted of a copper thimble holding a powder charge. The breech end of this was covered by a thin blue paper wad, while the bullet end had a central circular hole $\frac{1}{8}$-inch wide in it; this fitted over a small neck or stud cast on the base of the bullet, which was swaged over inside the copper thimble in order to carry it with it. The principle was not satisfactory, as there was a tendency for the copper thimble less its base to remain wedged in the chamber, preventing reloading till it had been scraped out. Muzzle loading was in general vogue until after 1860-1865, by which time breech loading in various forms of cartridge was taking precedence in popularity.

The needle-gun cartridge (1842), with a cap at the base of the bullet and exploded by a needle striker that penetrated the whole length of the powder charge, was one of the first types of cartridge. The shell was made of self-consuming linen or paper, and burnt and blew out with the charge. The gas escape at the breech made this system dangerous and unpleasant, but pistols for this cartridge—the Farington in 1865 and the Nestler of 1867—have been evolved.

Europe was behindhand with her official adoption of the metallic breech-loading cartridge, which, although invented in the shape of the Flobert pin-fire and the Lefaucheux pin-fire in 1851 and embodied by Lancaster in a metallic *centre-fire* cartridge in 1852, did not obtain recognition till many years later, or rather till the genius of American arms companies and private gunsmiths of all nationalities had produced it to such a state of efficiency that the "official" weapons and ammunition systems were hopelessly obsolete. The prevailing military fashions in needle-guns were probably responsible, but the earlier all-metal central and rim-fire cartridges were by no means as well made or as cheap as their descendants of the late nineties and the twentieth century. Specimens of early pistol cartridges in the author's collection show a wonderful crudity of manufacture—ill-fitting caps and cap-washers, clumsy bullets and swage marks on the brass case, all combining with a thick and clumsy rim to produce a cartridge of dubious efficiency, prone to jam from "proud cap" (*i.e.*, the cap blowing out or back on to the striker) and difficult to extract when fixed.

The rim-fire bulleted breech-cap of ·22 calibre soon evolved into the ·22 short, ·25, ·300, 320 long and short, ·380, ·410 short and long, ·442 —the latter the first official revolver cartridge used in the Enfield revolver as an arm for the British Service, and a variety of long ·44 calibre cartridges for Remington, Colt, Frank-Wesson, Forehand and Wadsworth, Marlin, and other American pistols and revolvers. The ·410 rim-fire short is to this day the favourite cartridge for Derringer pistols; and the modern ·22 cartridge, in all its variations and lengths, charges and special adaptions, is evolved from the original and still popular No. 1 bulleted breech-cap shown by Flobert over fifty years ago.

The rim-fire principle was simple, and consisted in spinning a copper alloy shell with a wide-flanged hollow rim. This rim and the base of the shell was charged with fulminate, which detonated when pinched between the face of the breech and the pin or striker of the hammer. Rim-fire cartridges cannot be reloaded, and are less certain in their actions than centre-fire. A point against them is, that in process of time the continual fall of the hammer causes a depression in the face of the breech-block into which the cartridge is forced by the striker, but without detonating. Rim-fire weapons should never be "clicked" on an empty chamber. They possessed an advantage over the pin-fire device in that they could be put into the chamber any way up, and were adaptable for repeating mechanisms. Rim-fire revolvers were common after 1870.

The pin-fire of Lefaucheux cartridge was, and is still, popular in France, Belgium, and on the Continent generally, and numerous cheap revolvers to take this ammunition are annually turned out. It has not changed from its original form, and is still made in ·6, ·7, ·9, and ·12 mm. The principle is simple. At the base of a rimless cylindrical cartridge-shell is set a thick wad, recessed to receive a small cap set vertically to receive a brass pin that projects externally and at right angles to the line of the chamber. These pins project beyond the cylinder, and receive the blow of the flat-nosed hammer which drives them into the cap exploding the charge.

All these systems have given place to the modern central-fire "all-metallic" cartridge-case, which, when properly made and loaded, is practically certain in action, invariable in effect, waterproof, and not likely to deteriorate much from storage. Very small matters—such as the effect of hot climates on the charge, minute variations in the load of the cap, swedging or crimping the shell or case on to the bullet, and similar points—make a wide difference for target work, but do not much affect the efficiency of the load for ordinary purposes.

By 1870 the central-fire cartridge-case for pistols and revolvers was in use, and by ten years later it was universal in all but cheap weapons and the ·22 and ·410 calibres. In the centre-fire cartridge the shell is drawn with a recessed base, in which is formed a perforated dome depressed toward the interior of the shell. Into this recess is forced from outside a cap containing a charge of fulminate and a small metal stamping called the "anvil." This abuts upon the dome, but leaves a clear space for the flame produced by the explosion, when the cap is crushed upon the anvil by the striker, to pass, and penetrating through the hole in the dome, fire the charge. Different makers have different shaped caps and anvils, and in America, where the cap is called a "primer," the anvil and cap are often self-contained, a process which simplifies amateur reloading. The modern central-fire cartridge-case is easily reloaded, but this is a practice seldom indulged in in this country. In the U.S.A., on the other hand, many people reload their own cartridges, cast their own bullets, and prepare their own ammunition. Broadly speaking, the waste of time is not worth the small saving in money, but for those who are interested in experimental loads, or who make a hobby of the loading, complete reloading outfits, powder measures, swages, moulds, etc., can be bought in the shape of most ingenious tools from the "Ideal" Company, but it is only fair to add that hand-loading very seldom approaches the all-round excellence and reliability of ordinary factory-loaded ammunition.

English cartridges differ from American in having the bullet and shell of the same external diameter, the heel, or portion of the bullet within the case, being reduced to admit of the sizing. Such cartridges are the ·320, the ·380, and ·450;

in each case the final "o" differentiates them from the American ·32, ·38, etc. In the American sizes the cartridge-shell is of greater diameter than the bullet, and the chamber is more acutely shouldered, the section admitting the shell being of greater bore than that taking the bullet. This principle is far better, as it prevents the bullet being deformed at that portion of its base most needed to prevent the escape of powder-gas past it, and most modern English revolvers are made for the American or S. and W. cartridges. In all modern central-fire revolver ammunition lubrication plays a large part, and the grease for this purpose is held in recessed rings, or cannelures, round the base of the bullet inside the cartridge-case. This, in all smokeless powder ammunition, is tightly crimped on to the bullet, in order to retain it sufficiently to insure its not creeping out of the case from the recoil of neighbouring chambers or commencing its travel before combustion of the smokeless charge (which needs a certain amount of confinement) is properly commenced.

CHAPTER IX

AMMUNITION (*continued*)

THE smallest modern central-fire pistol cartridge is the ·25 rimless automatic, used in automatic pistols and some Continental pocket revolvers. It is of necessity a smokeless powder, high-velocity load, with nickel-covered bullet either completely encased (full metal patched) or soft nosed—that is to say, with the lead core protruding at the nose, a practice that causes the bullet to expand upon hitting an object, producing a larger wound. Most metal-patched bullets are provided in both these forms.

Next comes the ·32 series, which are adapted for pocket arms—ladies' use—and the smaller calibre police weapons. Of these, the ·320 or ·32 Short Colt contains 9 grains of powder and a bullet of 80 grains, the popular ·32 S. and W. 9 grains powder and 85 grains bullet, while the ·32 Long Colt is 12 grains powder and 90 grains bullet, and the ·32 Colt Police Positive and ·32 S. and W. Long are each of 13 grains powder and 98 grains bullet.

The short cartridges are useful for cheap target practice, and give practically no recoil, while the long ones are suitable for longer ranges and more serious work. The increased powder charge, however, makes the recoil of a light weapon noticeable.

The popular ·32 automatic of 74 grains bullet, and the military automatic cartridges for the Mauser and Luger-Parabellum, also belong to this series, as does the ·32-·20 W.C.F. rifle cartridge, used in some police and military revolvers. This latter cartridge, loaded with the equivalent of 20 grains black powder, is smokeless, and the metal-patched bullet is capable of exceedingly accurate work.

The odd ·32-·44 and similar gallery charges are now seldom used except for trick work, and are not readily attainable unless ordered in quantities direct from the makers.

The ·38 series starts with the standard ·38 S. and W. of 14 grains powder and 145 grains bullet, suitable to most ·38 pocket revolvers. The ·38 Colt Short, 18 grains powder and 130 grains bullet, Long Colt D.A., with 19 grains powder and 150 grains bullet, and the regulation U.S.A. military cartridge. The ·38 S. and W. Special, 21 grains powder, 158 grains bullet, is probably the best cartridge of this calibre yet designed. The ·380 Automatic, with 95 grains bullet, is the pocket weapon size ; the military models of ·38 Auto, 130 grains bullet, and 9 mm. Luger, of 125 grains bullet, and the ·38-·40 W.R.F., 40 grains of powder, 180 grains lead bullet, are the more powerful military sizes.

The ·38 Special is the best target cartridge ; the ·380 Auto the largest pocket automatic size ; but the bulk of ·38 revolvers are pocket weapons, chambered for the ever-popular ·38 S. and W. short, a cartridge that has never been surpassed for its purpose.

The ·41 sizes are now all but obsolete, as are the ·44 Webley and ·44 Colt. In the latter calibre the ·44 Russian was the popular load for the old break-action Smith and Wessons, but the modern models are now chambered for the ·44 S. and W. Special of 26 grains powder, 246 grains bullet. The ·44 Russian is still used in the ·44 Colt, and is loaded with 23 grains powder, 246 grains bullet. The ·44-·40 W.C.F., 40 grains powder, 200 grains bullet, was long a favourite revolver cartridge, as it was used in the Winchester, Marlin, and Colt rifles and carbines; but the most celebrated of all revolver cartridges is the ·45 Colt, 38 grains powder, 255 grains bullet. This was the load with which most of the world-famed Western revolver work was done, and it is still the premier favourite with ranchers, miners, engineers, etc.

What the ·45 Colt is to the Western Hemisphere, the ·450-·455 British—the English official revolver cartridge—is to Europe, Asia, Africa, and Australasia. The short ·450 is now almost obsolete, as its charge of 13 grains powder, 226 grains lead, was insufficient for stopping purposes. The modern ·455-·450, with cordite equivalent of 20 grains powder and 230 grains lead, is an ideal cartridge, and is the standard Bisley load.

In the automatic class there are only three ·45 calibre cartridges—the ·45 Auto Colt, 200 grains bullet; the ·45 Auto Colt, U.S.A. Government Model, with 230 grains bullet; and the

·455 Automatic Webley, British Service Model, 220 grains bullet.

In the days of black powder, the fouling produced by the charge was such that, for accurate target shooting, the weapon had to be cleaned after each discharge of six shots. With modern smokeless powder all this is done away with, and shooting is a cleaner and pleasanter recreation.

Care should be taken not to use smokeless powder cartridges in weapons designed for black. The ·38 S. and W. pocket revolver can be used with smokeless, but the larger calibres, with tip-up or break-down action, were barred at Bisley as likely to blow open. All modern Webleys, Colts, and solid-frame Smith and Wessons, are designed for smokeless powder, and no automatic is made for black. For ladies' shooting or restricted practice, where noise is undesirable, the ·22 is the best calibre; the ·32 for house or small pocket weapons; and the ·38 or ·380 for large-bore pocket weapons for self-defence.

The only target size used in England is the ·455, which is also the military cartridge, and it is the only one allowed in the majority of revolver competitions at Bisley. A special flat-nosed target bullet, giving additional bearing surface in the barrel, is the most popular make; and the Webley hollow-fronted, man-stopping bullet is the best military or self-defence revolver projectile ever marketed. On hitting flesh the hollow front mushrooms, and develops all the shock of the impact within the body.

Automatic pistol cartridges are rimless, and

must never be used in revolvers or pistols for which they are not designed, even though they are similar in appearance and fit the chamber of the weapon in question.

Revolver cartridges in an automatic will jam the magazine, and do not function well in the barrel.

CHAPTER X

MODERN MILITARY ARMS

SINCE the invention of the revolver it has invariably been regarded as a high-class military weapon, and the legitimate successor to the pair of pistols which formed part of an officer's equipment ever since the sixteenth century. Nowadays the military weapon is the standard type from which all other variations of size, design, or calibre are taken. All officers of modern armies are equipped with one or other of the standard makes of military revolver or automatic pistol, and in some branches of the Services these weapons are issued to the rank and file as well as non-commissioned ranks in addition. The U.S.A. cavalry carry the pistol as a side-arm, and British artillery drivers, Army Service Corps, and non-commissioned officers of many branches, are also equipped with them, as are now the Motor Cyclist despatch riders of the R.E. Special Reserve and Cyclist Battalions, and certain machine-gun crews. The U.S.A. Army and Navy were equipped with ·38 "Special" calibre Colt's or Smith and Wesson's revolvers, and the ·45 U.S. Service automatic Colt is now being issued. The British Army uses the Webley Mark IV. or Service Model ·455 calibre revolver, while the Navy is now equipped with the ·455 calibre Webley-Scott automatic pistol.

The French Army uses a special revolver, with solid frame and extraction similar to the Colt and Smith and Wesson systems; but the cylinder swings out to the left instead of to the right, and is chambered for a special copper-coated bullet, high velocity smokeless cartridge of 9 mm.

The German Army uses the Luger or Parabellum automatic, and the Russians employ the Nagant—a revolver patented in 1895, and fitted with a wedge action, holding the cylinder close up against the rear of the barrel to avoid gas escape. Various other countries use the 9 mm. Browning military model, the Mannlicher, and the Bayard automatics as standard arms.

The military revolver is always the model to set the standard, and the little pocket revolvers, or special target models, are merely reductions or variations from standard types.

After the 1890 period of development had passed, Colt's set to work to perfect a self-ejecting model to compete against the Webley and Smith and Wesson types. They selected the solid frame swing-out cylinder model, with hand-operated simultaneous ejector; this type of arm has the benefit of greater rigidity of frame than any of the "breakdown" or "tip-up" action weapons, and further maintains the cylinder in better alignment with the barrel. This standard model of Colt revolver is now made in a complete range of barrel lengths, calibres and sizes ranging from the ·45 New Service Army model through the ·38 special military and police calibres into pocket weapons and even the ·22 target revolvers. There are three distinctive frame sizes—the New Service, or largest, bored for the ·45 Colt; the ·44-·40, ·44 Russian; ·38-·40 and ·450-·455 British cartridges. The "Army Special" or "Officer's Model" of medium size and weight, made for the ·41 long Colt, ·38 S.W. Special (the U.S.A. Army cartridge) the ·38 long Colt and the ·32-·20 follows. Next comes the "Police Positive" frame, with a variation of calibres from ·38 Special, ·380 Colt, ·32-·20, ·32 S.W., and ·32 long Colt. The ·22 Police Positive Target revolver is the only rim-fire weapon, with the exception of the Colt Derringer pistol, that is made, and with the exception of this last all the above weapons are double action.

In 1902 the Smith and Wesson firm issued their first solid-frame military arm, in which the familiar breakdown action was abandoned. All the previous models, with the exception of the S.W. Hammerless New Departure Pocket Revolver, were thereby rendered obsolete—a class which includes the ·44 S.W. Russian model single-action target revolver, the double-action ·44-·40 and ·44 Russian, and all large size breakdown action Smith and Wesson revolvers.

In 1905 the shape of the stock was changed to a square-shouldered pattern, and in 1908 the big ·44 solid-frame S.W. target model revolver was issued.

The Smith and Wesson solid-frame revolver is similar in design to the Colt, the same principle of double action and simultaneous ejection by means of the swing-out cylinder being employed. The lock work of these revolvers is noteworthy, mostly spiral springs being employed, and the exquisite internal finish and smoothness of action being a triumph of mechanical perfection. The cylinder of these models revolves from right to left in place of the usual reverse motion, and in the large calibre model a special front locking device is incorporated in front of the cylinder, with a view to increasing the rigidity of the cylinder bearing. A range of frame sizes and calibres to take the ·44-·40, ·455 British, ·38 Special, ·38 Colt, ·32-·20 W. Colt, and ·32 S.W. long cartridges are made, the ·455, ·44, and ·38 Special being standard military sizes and obtainable in the target model.

The Webley design of frame, with the safety stirrup locking bolt for the top strap, has always been so strong and well made that no need for a solid-frame action was ever manifested, even when the heaviest charge of smokeless were used in the weapon. The series of Webleys ran up to the Service model ·455 Mark IV., with 4 or 6 inch barrel; the 6-inch Army model, with curved "parrot beak" handle, still the favourite Army officer's weapon; the exquisite 1892 W.G. target model, with special grip, 7½-inch barrel, and target sights; and the later improved model ·455 6-inch barrelled Webley-Scott, in which a shorter hammer travel and different shaped grip more useful for cavalry use were adopted. All these weapons were ·455 calibre for the Service cartridge, but the firm also issue a range of miniature models for pocket and police use in the ·38, ·380, and ·320 calibres, the latter weapon being made hammerless, or with a special flush-hammer and small holding-down barrel catch for pocket use.

This summarizes the list of the modern revolvers

made by the three great arms firms, and in later chapters the special weapons suitable for special purposes will be indicated. As to the differences between them as shooting weapons it is impossible to judge. They are all of excellent design and manufacture, and it is purely a question of personal predilection, one man choosing one make, one another; though for quick double-action work there is no doubt that the Webley is superior to the other two, while their target efficiency for single-action work is about even.

The chief requisites of a military pistol or revolver are a capacity for hard wear under adverse conditions (for upon Service they may have to go for long periods without cleaning or attention), sufficient strength to resist without damage such accidents as being dropped, and a complete inter-changeability of parts to facilitate repairs.

The stopping power of the weapon is the point of most importance, and it is universally admitted that calibres smaller than ·45, whatever the velocity of the bullet, do not act satisfactorily as "man stoppers." Most modern Continental automatic pistols err in this respect, and it must be remembered that a man in action, unless hit in a vital spot, such as the brain or heart, when the effect is instantaneous, has still sufficient energy left to be able to close with and kill his opponent with a bayonet, despite many small-bore rifle and revolver wounds upon him. The effect of a revolver bullet should be to develop as much shock as possible when it hits its target. The tendency of the nickle-covered small calibre high velocity projectile is to go clean through a man and travel for a long distance the other side, while the soft lead bullet of the big bore ·45 revolver develops and utilizes most of its energy inside the body of the man hit.

A good instance of the value of large-bore lead bullets as against high velocity automatics occurred a few years ago in Mexico when some peons attacked two white engineers. One—a German—was armed with a Parabellum; he shot his adversary four times, each shot being in a vital spot, but the peon closed with him and ripped him up with a machete. The other man was armed with a ·44-·40 Colt Frontier; his first shot hit his adversary in the shoulder, and the shock was sufficient to knock the man down, allowing him his reserve five chambers to devote to the rest of the gang.

A reasonable objection to automatics as military weapons is their tendency to jam when dirty, an objection from which the revolver is reasonably free. Another most important point is that of mis-fires. In a revolver a mis-fire is of slight effect, as the next chamber can be fired within half a second; but in the automatic several seconds must elapse, and both hands be employed to "break" the action and eject the faulty shell.

The first official military revolver used in England was the Enfield of ·442 calibre, later made in ·450. This pistol was of the breakdown type, and automatically ejected the cartridges as the cylinder slid forward along its rigid axis. This weapon gave place to the Webley self-ejector Mark I. of ·476-·455 calibre, the long

·476 cartridge having a bullet with a deep recess at the base, in which fitted a hard clay plug, which expanded the base of the bullet to take the rifling. Officers have to supply their own revolvers, and any weapon of approved pattern and ·455 calibres, to take the Service cartridge, is accepted.

Solid frame pistols of the long-barrelled R.I.C. model, Tranter's or Adam's, were popular about the 1885 period, but the short ·450 cartridge proved insufficient to stop the Soudanese "Fuzzy Wuzzy," and among many officers double-barrel ·20 bore pistols, firing either a charge of slugs or the short ·577 Snider size cartridge, were popular.

Later the ·455-·476 cartridge was developed and adopted as standard, and Colt D.A. Frontier models were as popular as Webley's. About this time Edwinson Green, of Cheltenham, developed a revolver of graceful lines, embodying the advantages of the Webley stirrup fastening and the grace of the Smith and Wesson, but, unfortunately, the manufacture of these arms was not proceeded with.

Occasional specimen weapons by Kynoch, Silver, and numerous makers are found, but from this time onward the three great firms of Colt, Smith and Wesson, and Webley are the only ones of any consequence.

The weapons now permissible in the Army are any of these makes that are double-action, of not more than six inches in barrel length and chambered for the ·455 cartridge with 6¼ grains of cordite. The first automatic pistol built for the British ·455 cartridge was the now obsolete Webley-Fosberry, a six-shot weapon of the ordinary revolver type, but in which the barrel and cylinder recoiled along guides, cocking the arm and rotating the cylinder to the next chamber. This weapon possessed the drawbacks of a revolver, and only operated as an automatic when held rigidly, for if fired with a slack arm, or, as is the case with many expert quick shots, the shooter was accustomed to allow the muscles of his arm to absorb part of the recoil instinctively, the arm refused to function. It was also higher in the hand than most revolvers, and consequently jumped unpleasantly.

The W.G. and Webley-Scott patterns of revolver differ but slightly—differently-shaped handles, cylinder release gear, and breech-block plate being fitted to the latter model, which has also a shorter trigger pull.

The Webley-Scott ·455 automatic issued to the Navy is chambered for a special rimless cartridge of 220 grains bullet, and is readily dismountable by hand. It holds eight cartridges in a magazine in the handle, and is equipped with a large safety-grip on the butt, rendering the weapon innocuous unless gripped in the hand. At the moment of firing barrel and breech-bolt are locked to together. When the two have travelled in this position a short distance to the rear, the barrel rises upon its diagonally situated guides, and the breech-block alone travels back, ejecting the empty case, cocking the hammer, and reloading the chamber. When empty the mechanism remains open till a new magazine is inserted, when pressure upon a button releases the breech-bolt which loads the chamber, leaving the weapon ready for immediate use.

The design of the Webley is serviceable but ugly, and the shape of the handle set as it is—almost at right angles to the barrel—is such as to render instinctive shooting with this weapon less easy than with the curved handle pistol or revolver, or the automatic with shaped or slanting handle of efficient modern design. It is, however, a powerful and exceedingly accurate shooting weapon.

The Colt ·45 military model is now also made to take the British Naval Service ·455 automatic cartridge, but as a military automatic has much more to recommend it than the Webley, the slanting grip and easy natural alignment of the weapon rendering it a splendidly efficient weapon for service use.

The safety devices are worthy of note, and the whole design is one that combines the improved mechanical efficiency of an automatic pistol, with an attempt to secure the natural shape of a pistol adapted to the needs of the anatomy of the human hand, a point of design that less progressive makers see fit to ignore.

The accuracy of automatics surpasses that of the best military revolvers, and the speed of firing (normally three to four shots per second) is three times that of a double-action revolver in the hands of an ordinary shot. The objection to the use of automatics for target work is that they have mostly fixed military sights fitted, and that the ammunition is much more expensive than that used in revolvers of equal calibre. The balance is usually poor, and the trigger-pull of a dragging, unpleasant quality, incapable of delicate adjustment. The worse makes have the line of sight too far above the hand, and the racket of mechanism and the loud report all conspire to breed a tendency to flinching. The novice should avoid automatics, as they are far harder to shoot with than revolvers, and it is far better to learn the elements of pistol shooting on a safe and simple weapon rather than the complicated and cumbrous "latest little thing in automatics" advanced by the astute gunshop salesman.

For use as an English Army pistol the "W.G." Officers' model, or the Webley-Scott Army revolver, according to which grip you prefer, can be cordially recommended. The New Army Colt is also an excellent weapon, as is the "Webley-Wilkinson," but as different makes suit different types of hand, it is always best to choose your pistol personally, rather than leaving it to a friend to select.

CHAPTER XI

TARGET WEAPONS

THE target revolver, as used in competitions at Bisley under the rules of the National Rifle Association, in the U.S.A. under the United States Revolver Association, and in other countries under similar national bodies, is usually of the standard military calibre of the country in question, and subject to definite rules as regards length of barrel, trigger-pull, adjustable sights, etc., unless where otherwise specified in the rules of the competitions.

The length of barrel for a target revolver for use at Bisley must not exceed 7½ inches, and the most popular models are the Colt "New Service" ·455 British 7½-inch barrel, the Smith and Wesson of similar type and calibre, and the Webley W.G. Target model, or the Webley-Scott with target sights and long barrel. The single-action S.W. Russian model and the Frontier and Bisley model Colts were for a long time favourites, but now all single-action revolvers are practically obsolete, the S.W. tip-up action of old pattern having been barred at Bisley as liable to blow open at the breech, under the strain of smokeless powder.

Modern conditions demand that the fore-sight

of a revolver shall be fixed and incapable of adjustment, but the back-sight may be capable of lateral movement. On military models neither sight may be movable, and the rear-sight usually consists of a plain slot or notch in the top strap of the frame, while the fore-sight is a plain section of metal, either milled integrally with, or set immovably into, the end of the barrel. Military fore-sights should be rounded, in order to insure freedom in drawing the weapon from a holster.

On Target models a bead or "Ira-Paine" fore-sight is popular, and such sights, when sent out from the makers, are cut with a tapering bead, so that any size of bead to suit the user can be made by filing the length of the sight to the requisite size. Many experts prefer a plain square fore-sight with a square notch in the back-sight in place of the bead and V-shaped rear notch, and back-sights of varying heights for different ranges are issued by the makers of target revolvers.

In general, it is best to select the sight that suits your own personal individuality, rather than to be persuaded that So-and-so's speciality is bound to suit you; as a hard-and-fast rule, the plainest sight that you can use is the best to adopt.

A great deal of refinement can be expended upon the adjustment of trigger-pull of target weapons. The official pull is a minimum of 4 lbs. for revolvers, but many private shots have weapons (unusable in competitions) with a 2 lb. trigger-pull. It is essential that there should be no "creep" or "drag" to the pull, and adjustments must be made by a skilled gunsmith, and should not be attempted by amateurs. If, however, this is done, it should be remembered that after the scear and hammer notch have been gently filed to nearly the suitable adjustment, finishing off is done by means of small triangular slips of stones of varying degrees of cutting surface; these leave the metal in a polished or burnished condition, and it is only by their use that variations of an ounce or so can be effected.

Outside the limits set by international competition usage come a wide variety of target models of various calibres, both pistols and revolvers, suitable for private use or "open events."

Among the revolvers suitable for this are the ·38 Officers' Model Target Colt, which is a replica of the larger New Service Target, and a splendidly accurate weapon. The ·32 Police Positive, with target length barrels 7½ inches or 6 inches, and target sights, make excellent weapons for the use of ladies, the light charge and low recoil rendering the weapon pleasant and comfortable to use. The ·22 Target Colt is chambered for the ·22 long rifle cartridge, and is the only first-class ·22 revolver of any use for target work at present on the market.

The Smith and Wesson models are also marketed in similar calibres and types, and a variety of cheap American pistols, fitted with speciously named "Wild and Woolly West Target Stocks," etc., are also to be bought. These weapons, while quite excellent for house defence and similar onerous duty, do not stand up to prolonged use, and are useless as target revolvers or for any duty where much shooting is expected of them. The cheap ·22 weapons are usually unsatisfactory and "keyhole" badly. To buy cheap revolvers is waste of money, as in order to get good results the best materials and workmanship are necessary; also, if you ever want a revolver you will want a reliable one. The thirty-shilling specimen is worthless as a competitor against the three-pound weapon of reputable make, and to buy the best is the only real economy. Cheap foreign pistols of the twelve and sixpenny variety are more dangerous to their users than to anybody else, and should be avoided at all costs.

Turning from the target revolver to the target pistol we find a range of ·22 calibre weapons by good makers, inexpensive to purchase and capable of the highest accuracy.

The Webley ·22 Target Pistol has a pleasant grip and balance, while the ·22 semi-automatic manufactured by the same firm is an ideal pistol for accustoming one's self to the grip of an automatic, and for general target or sporting use. The new Colt ·22 Target Automatic is also excellent.

The Smith and Wesson single shot pistol is made in a range of calibres from ·22 to ·38 rim-fire, as are many other American pistols; but as all cartridges, with the exception of the ·22 in variations up to the "long rifle" size, are hard to get in England from the average gunsmith, unless specially ordered, and quite unobtainable in most parts of the globe, the ·22 calibres are the only ones worth considering for the average user. The latest model ·22 S. and W. has a double action, the only target pistol so equipped.

The Stevens pistols of ·22 calibre are made in varying weights and several models, called "Diamond," "Gould," and "Conlin," of which the "Gould" is the most popular with target shooters. The action is simple, and on the tip-up system, with an automatic extractor to push out the cartridge when the barrel is depressed. The only difference between the models is the weight and shape of the grip. The Smith and Wesson pistols, like their revolvers, have the same hinge tip-up action extractor and top fastening, while the Webley target pistol and most Continental weapons have a kick ejector of the hinge type. The Webley semi-automatic is unusual in design. Built on very similar lines to the Webley ·32 pocket automatic, but without the magazine and recoil springs to close the action, the weapon, when fired, blows open, ejecting the empty shell and cocking the hammer for the next shot. The breech is left open for reloading, and the user has only to put in another cartridge and close the sliding breech to be ready for the next shot. An additional extractor on the barrel is provided for use in case of a broken-rimmed cartridge, or the mechanism not functioning properly. This pistol can also be fitted with a detachable shoulder stock, and forms, when so fitted, an exceedingly convenient portable rifle.

The German, Belgian, and French makers turn out some wonderfully complicated ·22 target pistols with adjustable sights and curiously-shaped crutch handles. These weapons are often fitted with hair triggers and a variety of automatic ejectors and other fitments; unfortunately, the soft materials of which they are made, and the typically bad workmanship of the Continental gunsmith, make them so delicate and unreliable as to render them practically worthless for ordinary use. When in going order extraordinarily accurate shooting can be made with them, and for trick work, like picking off the pips of the ace of clubs, they are invaluable.

The ·22 cartridge is exceedingly accurate, and is the only size that it is safe to use in such a thickly populated country as England, unless shooting is done upon a properly constructed range. Up to 50 yards the ·22 short gives excellent results, but for longer distances, either the ·22 long or ·22 "long rifle" should be used. These give accurate results at 100 yards and over. In target revolvers of this calibre the ·22 Long Rifle "Armory" model should be used, as the bullet is in this type crimped into the shell. If the uncrimped varieties are used in a revolver, there is a tendency for the recoil to loosen the bullets in the adjoining chambers, with the result that their noses project and prevent the cylinder from revolving for the next shot. In all ·22 weapons smokeless powder should be used, and it should also be remembered that the corrosive effects of the gases of heavy fulminate of mercury charge in the rim are greater than is usual with centre-fire cartridges, so the weapons should be cleaned with cordite cleanser, as well as ordinary oil.

For the shot who wishes to compete at Bisley a ·455 calibre revolver of approved pattern is essential, but the man who wants a target pistol for use in his own grounds and for his own amusement rather than for serious competition work can be confidently advised to invest in a good ·22 revolver or pistol with its cheap ammunition, exceptional accuracy, and greater margin of safety for the neighbourhood. Above all things, avoid the "special revolvers," chambered for odd gallery-sized cartridges, such as the ·38-·44 Smith and Wesson that have been so much praised in various works. The reason is simple—though the pistol may be excellent, the cartridges are practically unobtainable, unless specially loaded in lots of 1,000 or more, expensive, and no better than the standard models.

For folk who prefer a heavier charge than the ·22 and like a revolver that can be used as a serviceable weapon for self-defence or travelling abroad, but who dislike the heavy military models, or for ladies who dislike the heavy recoil of the ·455, the range of target revolvers of ·38 and ·32 with 6-inch barrels should easily provide an ideal all-round weapon.

CHAPTER XII

THE AUTOMATIC PISTOL

THE greatest modern development in the design of firearms is undoubtedly that of the automatic principle.

In this, the recoil of the weapon is utilized to function the operations of ejecting the empty shell, cocking the arm for the next shot, and reloading the chamber from the magazine. Two distinct systems of automaticity have been used. The first is the operation of the arm by means of gas pressure tapped from the barrel near the muzzle. The second—and up to date the only one practically applied to pistols—the utilization of recoil.

The automatic principle was first applied in machine guns, all of which were gas operated, and later the recoil principle was applied to firearms for hand use and especially to pistols.

For a long time development was delayed by the existence of master-patents that prevented the utilization of various basic ideas and their adaption to the modern cartridge ; but upon the expiry of the master-patents many small inventors and several great arms firms set to work to develop the automatic principle as applied to firearms in general, and pistols in particular.

The first automatic pistol to appear in the

market was the Borchardt (1893), the predecessor of the present Luger-Parabellum and employing the same base idea of a toggle-joint breech. The magazine was concealed in the handle and the pistol was usually equipped with a skeleton stock for use in converting it to a short carbine. The action was unsatisfactory and usually jammed. This was in turn succeeded a year later by the Bergmann, a weapon that can claim to be the ancestor of all flat pocket-sized automatics. In this pistol cartridges were loaded by means of a clip into a magazine placed forward of the trigger-guard, whence they were forced by a spring-actuated under-lever into line with the breech-bolt. The arm was well designed, but the long travel of the breech-bolt and the weakness of the spiral spring that returned it to position frequently caused it to jam. This weapon and its imitators, such as the "Simplex," suffered from the uneven quality of the ammunition then made for them, as smokeless small-calibre revolver cartridges were at the time a novelty and by no means so reliable or evenly loaded as at the present time.

The next weapon to appear was the Mauser of 7·65 calibre, the 1898 model combination automatic pistol or carbine, the wooden holster serving as a stock attachment. This was the first really satisfactory automatic pistol to be marketed, and the large-bore model of 9 mm. that is in use to the present day is practically the same weapon. In this piece the barrel and breech-block recoiled together for a short distance, after which the automatic check released the bolt and allowed it to continue its travel, ejecting, cocking, and reloading. The magazine was of the box type, set forward of the trigger-guard, and held eight cartridges loaded into it from above by means of a clip, which was held in a notch in the frame. On withdrawing the empty clip the bolt closed, automatically reloading the chamber. The lock work was remarkable, for the whole weapon could be dismounted without the use of tools other than the point of a cartridge, and did not contain a single screw, all pieces being milled and interchangeable.

The powerful bottle-necked cartridge and nickel-covered bullet were capable of good work up to 600 yards, and the weapon was sent out with a military standard sight graduated to 1,000 metres. This was for use as a carbine, the wooden holster-case serving as a stock. Viewed as a pistol simply, the weight, bad balance, and powerful recoil did not make it a valuable weapon as a pistol; but as a high-speed weapon for self-defence it was an advance on all its predecessors, and a number were used by the Boers during the Boer War.

It is interesting to note that it was with these weapons that two Russian anarchists were able to defy the police of London and initiate the memorable "siege of Sidney Street"—an affair which necessitated the calling out of the troops and raised such a popular outcry over the inefficient armament of the London police that they, too, were shortly armed with an automatic pistol of approved type.

Close upon the heels of the 1898 Mauser came, their rivals the Mannlicher firm, with the Mannlicher automatic, a weapon very similar to the Mauser, but with the magazine inside the handle, a type first started by the Borchardt, and popular ever since.

There were also many experimental weapons of lesser importance, few of which reached the outside world; the "Mars," a weapon made by a British firm, is an instance of this type.

By 1898 the name of the Browning automatic system began to be heard, and soon the Browning automatic pistol of ·32 calibre made its appearance, and owing to its excellence sprung into universal popularity as the latest thing in pocket pistols. The ·32 or 7·65 mm. Browning cartridge is now standard all over the world, and was further popularized by the adoption of the Browning patents by the Colt Firearms Co. whose excellent plant and material produced, in 1903, the ·32 Colt hammerless automatic, a weapon then much superior to the Continental-made Browning. The Browning pistol was soon produced in 9 mm. or ·380 calibre, and in 6·35 mm. or ·25, for waistcoat pocket use.

The Colt ·38 military model automatic had the barrel fixed to a hinged slide, so that breech and barrel were locked at the moment of firing, and was the first automatic pistol to be adopted as a military arm, and was soon made in various barrel lengths to suit varying requirements. The little ·25 calibre waistcoat pocket Colt automatic is practically a miniature of the ·32, but has a different firing mechanism, the hammer and firing-pin of the ·32 being replaced by a spring striker similar to that in use in military bolt-action rifles.

The next model was the ·450 Colt automatic, which was shortly succeeded by the ·450 new model automatic military model, especially designed for the U.S. Government, and embodied several radical improvements, notably the abandonment of the slide action of the earlier ·45 and ·38 military models in favour of the improved locking device for breech and barrel, and the embodiment of a new and comfortable sloping handle, and several safety devices.

The Webley automatic pistol of ·32 calibre was the first to succeed the hybrid Webley and Fosberry (see chapter on "Military Arms"), and was the first automatic pistol (not revolver) to be made by them. The principle employed is a variation from all other automatic designs, for no spiral recoil-springs are employed, the functioning of the breech-bolt being attained by the absorption of recoil by a flat laterally placed recoil-spring and lever situated in a recess in the right-hand grip at the side of the handle magazine. In this and the pocket models of ·25 calibre, as well as in the new Metropolitan Police model of ·32 calibre (which has a raised backsight in place of the plain notch, and remains open when the magazine is exhausted), the breech-block is not locked to the barrel at the instant of explosion, but in the high-power ·380, ·38, and ·455 military models breech and barrel are locked during the instant of firing. The Webley is the only British automatic pistol manufactured. Spiral springs are now used in the latest model hammerless ·25 automatic made by them, the recoil-springs running parallel to the axis of the barrel in the recessed breech-bolt.

The Savage Co. make an excellent ten-shot ·32 and ·380 pocket pistol which is remarkable for lightness, portability, magazine capacity, and excellence of manufacture. Their lock is dismountable without the use of tools, and the whole weapon is a piece of masterly design. The barrel and breech are locked at the instant of firing, and the barrel has to revolve in its seating for one-eighth of a turn before the breech is free to recoil. The magazine is contained in the handle, and the cartridges are "staggered," allowing a greater capacity than in any other automatic arm.

Numerous Continental makers have invaded the market with cheap automatic pistols of ·25 and ·33 calibre; among the best of these are the Bayard models, which comprise the smallest large-bore pocket automatics made.

Other makes are those of Pieper, Jeffeco, Victoria, Express, Walman, Roth-Sauer, Clement, and numerous others, but purchasers should remember that cheap pistols are false economy, and that where the weapon of a first-class firm like Colt, Webley and Scott, Smith and Wesson, F.N. (Browning), Mauser, or Mannlicher can be relied upon, the cheaper makes have no such warranty. *At its best the functioning of an automatic pistol is less reliable than that of a revolver, and the delay consequent upon a misfire or jam is such as to seriously prejudice them in favour of the revolver, in the point of view of anyone who desired a reliable self-defence weapon.* The author has personally tried almost every make and calibre of automatic pistol, whether British, American, or Continental, with the result that so far he has not come across the weapon that will not jam upon occasion. The smaller calibres are much the worst offenders in this respect, and it is usually urged in defence by manufacturers that unsuitable ammunition has been used. Despite this ingenuous disclaimer, a good automatic should, like a good revolver, be capable of use with the standard cartridge issued for it by any ammunition firm of repute. As a matter of fact, most jams in automatics are due to magazine trouble; that is to say, either the feeder end of the magazine has got slightly deformed, or else the cartridge-carrier platform and spring need attention. As a military arm for Government issue, where the user can always get a spare interchangeable part from his regimental armourer, the heavy-calibre automatics may be considered efficient, but the opinion of most private individuals who have occasion to carry a pistol, such as explorers, mining engineers, war correspondents, and the like, is that the large-calibre modern revolver is still the superior and more serviceable weapon, a conclusion with which the author, at pain of being thought ultra-conservative, yet heartily endorses.

At the moment of going to press the long-expected ·22 Colt Automatic Target model has made its first appearance in England, but so far only a few have been imported, owing to the dislocation of traffic caused by the war. The pistol is a radical departure from previous Colt designs, and seems to be in every way an ideal weapon. The magazine is designed to hold ten shots, and is housed in the stock. The weapon is hammerless, adjustable sighted, and has a removable breech side, allowing the barrel to be cleaned and inspected from the rear. The balance is good, and the length of barrel 6½ inches—ample for accurate shooting. The grip is set at a suitable angle to the hand, and the weapon "comes on" in an excellent line of sight. It is not necessary to use special cartridges: the ordinary long rifle rim-fire with greased bullet (or even the greaseless if the shells are slightly lubricated) function the arm with excellent results.

The weapon should prove enormously popular, as it is a type that has been long awaited by the shooting public, and uses reliable and inexpensive ammunition of limited range.

CHAPTER XIII

THE POCKET AUTOMATIC AND REVOLVER

THE bulk of revolvers and automatic pistols are made in sizes small enough to be carried in the pocket—powerful weapons, but easily concealed and portable.

The average Englishman has little idea how widespread is the habit of carrying a pocket pistol; it is certainly unusual in England, but almost customary in the U.S.A. and Continental countries. A brief inquiry into the sales by English gunsmiths of pocket pistols to English folk not travelling abroad would probably astonish the casual inquirer. After all, it is better to be efficiently armed than not; and ever since the first prehistoric man invented the sling or the bow, weapons that kept the enemy at a distance and disqualified mere personal strength have been popular.

For the ordinary man a pocket pistol is not an essential; but for anyone whose work or inclinations leads him to strange places in our own cities, or even as far afield as Paris, a pistol is a wise precaution, as it insures that, if you are molested, you hold equal or superior cards to your adversaries. Strangely enough, the English "rough" is not sufficiently familiar with firearms to be afraid of them, and is more or less certain that you will not dare to use them. A display of cold steel, or, for choice, a sword-stick, on the other hand, scares him off at once.

In Paris the indiscriminate *apache* uses fire-arms with absolute nonchalance, and street battles between rival bands are as casual as were similar affairs between Bowery or Chicago "toughs" or "gunmen"; as a result, wherever the criminal classes use weapons, the respectable classes also adopt them for private self-defence. The police all the world over share the trait of never being on the scene when wanted.

One cannot deal with the moral question of whether shooting, or even killing, in self-defence is allowable; but in defence of the matter of killing, it must be remembered that, when one is in personal danger and the victim of sudden attack, shooting is probably instinctive, without the least fraction of time wasted in sighting. The pistol is drawn and fired at a speed and during a period of nervous tension that probably entirely inhibits all but a purely instinctive mental process, and one shoots as one points or throws a ball—instinctively. There is only one rule to be remembered, and that is—shoot first.

The "shooting through the pocket" myth is dangerous, for even the expert shot cannot depend on getting his man in this way unless he is within three or four yards; and the author discovered by personal experience that an automatic discharged through the pocket is capable of one shot only Having occasion to scare a rather ugly customer, I fired a ·32 automatic of first-class American manufacture straight through my coat-pocket into the earth. The empty case was prevented by the enclosing pocket from ejecting clear of the action and jammed the weapon, putting me out of action. Luckily, no second shot was needed.

The shooting through the pocket idea was certainly useful in the days of short Derringers and quarrels with gamblers across a table; also in the days of the "southern gentleman," who wore a long black tail-coat, with a pistol in each tail-pocket, as the sweep up of the tails and pistol allowed him to fire from the hip with comparative freedom of action. The modern lounge suit is practically impossible for the purpose, except in the case of a *corps-à-corps* struggle. If you have time to draw a pistol, and your enemy has not "got the drop" on you, draw and shoot. You will have a six-to-one better chance than shooting at random through your suit.

In the old days "out West" the double-action, or self-cocker, revolver was condemned by experts, "bad-men," and sheriffs alike. They claimed that with it the average man, drawing hastily in a quarrel, sent the first bullet into the floor by drawing the trigger out of sheer flurry and nervousness. The action of cocking a single-action pistol took just long enough to steady one's nerve for the shot.

There is a good deal of wisdom in their summing-up, and it is a point that is accentuated by the bad design of most modern automatics. Pick up the average revolver or target-pistol and snap without sighting at a target, and you will probably find when you glance along the sights that you had instinctively pointed more or less true with it. Try this with any square-handled automatic, and you find that your bullet would hit the ground about ten yards in front of you. This is because the revolver handle is curved to suit the natural grip of the hand, whereas the square shape of the automatic necessitates a strained unnatural position of the hand, the wrist having to be much more "bent upward" to bring the barrel into true alignment with extended arm and eye. Practice will cure this; but the unpractised shot, shooting instinctively with an automatic, is much less dangerous than the same man with a revolver or curve-handled pistol.

The advantages of the pocket automatic are—rapidity of fire, and flatness and compactness in the pocket; the disadvantages—unnatural alignment, tendency to jam, and the danger common to all magazine pistols of taking out the charged magazine and forgetting the live cartridge in the chamber.

The pocket revolver is heavier, clumsier, and holds fewer cartridges than the automatic, but it is, owing to the projecting handle, easier to draw and easier to shoot with.

In making up one's mind over the choice of a pocket weapon, it is well to remember that if the arm is merely to be carried with only remote chances of needing it, an automatic is excellent; but if it is for probable use, and as a permanent article of wear, regarded as more important than a collar or tie in the country or circles in which you propose to travel, the revolver is much better for your purpose.

The Colt hammerless ·25, ·32, and ·380 automatic pocket pistols are good, and are equipped with a safety-grip in the handle, which prevents the weapon being discharged unless safely held in the hand. Falls or accidental knocks cannot discharge the weapon. The same applies to the Browning's or F.N. pistols of latest design and similar calibres.

The Webley ·25, ·32, and ·380 automatics are also sound pocket weapons, but all except the new ·25 have the disadvantage of an outside hammer, and are slightly less convenient in the pocket. The ·32 M.P. Webley has been adopted by the London Metropolitan Police as their official weapon.

The Savage ·32 and ·380 ten-shot automatic has many better points than most pocket weapons, including a very good grip, greater magazine capacity, and light weight; its shooting is also good.

The pocket Mauser of ·25 calibre is, like all the weapons of this firm, well-made and reliable, but for its small calibre is unduly large for the pocket. The Bayard, on the other hand, is little larger than the average ·25 pistol, and is made in ·32 and ·380 calibres, and is the smallest large-bore automatic on the market.

The Smith and Wesson ·35 automatic is rather similar in external appearance to the Webley, but the special cartridge necessary renders it less useful than other makes for standard ammunition that can be got all over the world.

The ·25 Victoria is an inexpensive Continental weapon, and is the smallest ·25 made; its mechanism is not well finished, but exceedingly simple, and one of this make in the author's possession has proved an exceedingly reliable little weapon.

On the whole, avoid cheap European automatics, and only buy first-class productions. The cheap stuff is unreliable, inaccurate, and even dangerous, while the saving of a few shillings does not warrant the possible accident that may result from their use.

In automatics, as with revolvers, the larger the calibre the more effective the arm, and though a ·22 or ·25 may be very pretty and portable, nothing under the ·32 automatic or the ·32-·20 W.R. revolver cartridge is useful as a stopping weapon, and of all sizes the ·380 or ·38 S. and W. is the best.

The reader may object that a pocket arm that will wound is amply sufficient for self-defence, but it must also be remembered that unless the wound is enough to disable the adversary it is of no use. The smaller calibres may wound a man, but not enough to embarrass him and prevent his killing you with a knife or "life preserver," while the big smashing effect of a large calibre bullet will immediately put him *hors de combat* without of necessity proving fatal. For a pocket self-defence pistol, neither heavy powder charge nor great velocity of the projectile have much value; it is preferable to use a heavy bullet and a small charge—a principle employed in the old-fashioned Derringer pistols, and proved by years of experience to be sound.

The pocket revolver should either be hammerless or devoid of projections to catch in the pocket or clothes when drawing the arm. The hammerless pistol has the additional advantage of being safe against accidental discharge, as it cannot fall upon its hammer and so fire the cartridge.

The best modern hammer revolvers are also fitted with an intercepting device between hammer and firing-pin or hammer-nose and cartridge; the Colt Positive Lock is a typical example.

The ·38 and ·32 Smith and Wesson hammerless safety revolver is one of the best pocket pistols of its type, as it embodies the grip safety with its hammerless feature. The ·32 Webley hammerless has also merits, but the safety device fitted does not act automatically—a point that may be overlooked, and leave the owner with a locked weapon in his hand, if he forgets to operate it when drawing the weapon.

Hammerless revolvers are essentially the safest of all firearms, as they are of necessity double-action. This involves a considerable exertion of pressure on the trigger to operate the mechanism, and renders the weapon fairly safe in the hands of children or people unused to firearms. Their disadvantage is, of course, that they cannot be used for single-action target practice, or anything but double-action shooting.

Hammer revolvers can be used either single or double action, but unless fitted with some intercepting device between firing-pin and cartridge are dangerous if dropped hammer downwards. Weapons not fitted with any safety-lock should always be carried with the hammer down upon an unloaded chamber, and never by any chance left at full cock.

The Webley Mark IV. ·38 is a compact and serviceable six-shot weapon—a miniature of the Service arm of ·455 calibre.

The Smith and Wesson ·38 Perfected is intermediate between their tip-up and solid-frame models, and though retaining the tip-up action and automatic ejection, has the handle and locking device of their solid-frame models. These are similar to their military models, and are made in various calibres and barrel lengths—·22, ·32 S. and W., ·32-·20 W.C.F., ·38 S. and W., and ·38 S. and W. Special. The 1902 model has a rounded stock, the 1905 a square. Numerous Spanish imitation Smith and Wessons are made, and though externally replicas of the genuine pistols, are distinctly to be avoided.

The Colt Pocket and Police Positive models vary only in the shape of grip and length of barrel, and the action is identical, the solid frame, swing-out cylinder, ejection and positive lock features being exactly similar. They can be got for the following cartridges: ·32 Colt, ·32 S. and W., ·32 Colt New Police, ·32-·20 W.C.F., ·38 Colt, and ·38 S. and W. Special. Several models take several different ·32 or ·38 cartridges of varying charge.

Among other pocket revolvers, the Continental "Velo-Dog" patterns are deserving of mention.

These pistols are made in all conceivable patterns of hammer, hammerless, self-ejecting, and plain models, to take a long centre-fire, high-velocity cartridge of ·6 mm. calibre, known as the Velo-Dog. The bullet is of lead, copper coated, and the velocity and penetration exceptionally high. The same type of cartridge is used in the 9 mm. (·380) French Service arm, and pocket revolvers for this cartridge are also common.

Many Continental pistols and revolvers are now made for the ·25 and ·32 rimless automatic cartridges; but these weapons, though outwardly well finished, are seldom well made inside, and almost invariably made of indifferent metal, badly fitted, and with terrific "pull-offs." Except for their small size, coupled with the powerful charge, they have little to recommend them.

The cheaper kinds of American revolver are, as a rule, efficient as pocket pistols, but will not stand up to much continuous use without wearing out speedily. They lack the smoothness of action and accuracy of the first-class makers, but are better than most Continental weapons, and are efficient and safe for house or self-defence arms.

The single-shot Colt ·410 rim-fire Derringer, and the double-barrel under-and-over Remington Derringer, are still made, and as large-bore, self-defence pistols for close work are preferable to many small-calibre repeating weapons. Although single action, and almost, if not quite, obsolete, they still have a steady market, and are now, as ever, the best weapons of their class. Many people prefer a single heavy-shot pistol to a small-bore repeater or revolver, and there is much to commend their choice; they are also the cheapest model of first-class pistol marketed.

In self-defence work, it is well to remember that an unloaded revolver is of less use as a weapon than half a brick in the end of a stocking, and that you should never draw a pistol unless prepared, if necessary, to shoot. Looking for a burglar with a lighted candle is as stupid as hunting for an escape of gas with the same illumination. It makes you into a target—not the burglar.

The law on the subject of shooting burglars is rather vague—technically he should put you in peril of your life (*i.e.*, have first shot); but, actually, a burglar is presumably an armed and dangerous criminal. Such a man, having been called on to surrender, presenting a harmless spanner, or, say, a pipe stem, or attempting to draw a concealed something from his pocket, could hardly expect one to wait to see if his implement was a real pistol and really loaded. On the other hand, a burglar shot in the back might lead to a lot of judicial unpleasantness.

It is wise to take advantage of all possible cover, and, except in countries like certain Latin-American Republics, where a wounded man means more legal trouble than a dead one, avoid killing if possible, but don't run risks yourself.

CHAPTER XIV

MISCELLANEOUS WEAPONS

THE exact limits of a book on pistols and revolvers are hard to define, but I think that the air-pistol, and especially the saloon-pistol, should be within its limits. A friend, seeing this loose manuscript with the above heading, was anxious that I should include details and data upon the Japanese sword and knife-throwing among the Latin-American races; as it happens, neither of these come within the scope of this work, and this chapter must deal with pistols that possess more or less vogue or merit, and by their novelty require some short mention.

The air-pistol has been known since a remote period, and was sternly forbidden as a weapon during the earliest days of firearms. A French book of 1783, entitled "Panoplie de Guerre," alludes to it as having first come into notice when one was presented by Martin de Lisieux to Henry of Navarre before 1600; and seventeenth and eighteenth-century compressed-air weapons of Nuremburg or Dutch manufacture occur as museum specimens. The earlier models were either charged by pump, or by exploding a loose charge of powder inside the air-chamber, and, once charged at full pressure, were capable of several shots before the pressure fell sufficiently to affect the accuracy of the shot. The facility that they afforded to assassins was always a restriction upon their manufacture, and it was not till about the 1850 period that the compressed-air weapon—notably as a walking-stick gun or "air-cane"—took definite shape.

A few pistols of this type were manufactured, and consisted of a steel butt or air-chamber, pumped to high pressure by means of a foot-pump, and fitted with a release-valve, the seating of which was made of buffalo horn. The barrel, lock, and stud-trigger were detachable from the breech, and some were fitted with a plain tap or cylinder breech-block for breech loading. The action was simple. The lock was cocked by means of a cocking-lever or key, and, upon pressure of the trigger, the internal hammer struck the projecting stem of the outlet valve of the compression reservoir. This, driven in by the blow, released a certain amount of compressed air, sufficient to discharge the ball, and closed again automatically. These weapons were made in smooth-bore and rifle sizes to take ball or shot, and were powerful enough to drive a ball through a ¾-inch plank at twenty paces. Later, the predecessor of the common air-pistol of to-day was evolved, and the principle of a released spring-driven plunger compressing the air in a chamber employed. These arms were made to take the No. 1 slug—a conical bullet of ·167 calibre, with a deep base cavity, but the weapons were little more than toys, feeble and inaccurate. The usual type depended upon the compression of air in a cylinder round the barrel, the barrel itself forming a hollow piston-rod connected to the plunger. To operate them the spring was first compressed by ramming the weapon against a wall or the ground, a screw plug at the rear of the barrel was undone, and a slug inserted which was pushed by a projection upon the breech-screw beyond the air-vent connecting cylinder and barrel. On discharging the arm, the spring released, drove the barrel and piston breech-block forward, compressing the air in the cylinder; this escaping into the barrel through the vent, expelled the slug. In actual practice the kick of the spring and the escape of air round the ill-fitting leather glands of the barrel and piston breech-block rendered the weapon inaccurate and feeble. A better type had a hollow-handle cylinder and bolt-action breech to the barrel. The spring was compressed by pulling down the spring-piston in the handle, which, when released by the pressing of the trigger, compressed the air.

The only modern air-pistol of any value as a weapon is the Westley Richards Highest Possible model, which is equipped with a rifled barrel and adjustable sights. In this pistol the break-action is employed, and the action of opening the arm compresses a spring and plunger-piston in a big cylinder under the barrel. When the slug is inserted the weapon is closed, and the base of the barrel forms an air-tight joint by means of a leather washer with the standing breech. On pulling the trigger the spring-plunger compresses the air in the cylinder, which, issuing through a port, passes through the hollowed breech-block and escapes into the barrel, propelling the slug. The arm is well made and comparatively accurate, but the shock of the discharged spring and the peculiar balance of the pistol render it useless for learning to shoot with a real firearm, and not much more than a toy. For indoor or garden practice, discouraging maurading cats, and similar work, it is efficient; but the penetration of the slug is not enough to bring down much larger game than possibly a pigeon.

An air-pistol slug in the eye would destroy sight and possibly kill, but otherwise the weapon is not fatal, though wounds of slight penetration can be effected with it, and it must decidedly not be classed as a "toy pistol."

Shooting or pointing at people should never be allowed, and children indulging in the practice, even with toy weapons, should be severely rebuked, and, if necessary, properly spanked. It may seem cruel to advocate "corporal punishment," even with a bedroom slipper, for such an essentially natural offence, but, as nine out of ten cases of "accidental shooting" arise from the pointing of arms at a person with the cheerful, jesting formula, "I will shoot you!" it is sound practice to discourage the habit from the earliest days.

The Pistols Act provides that no pistol or revolver can be sold to any person without the production of a gun or game licence. Pistol, within the meaning of the Act, means any air-pistol or firearm with a barrel less than nine inches long. Exceptions are made with regard to antique firearms—a rather vague classification, but which certainly covers flint-lock and probably all muzzle-loading weapons. Rim and pin-fire breech-loaders would not be technically "antiques." The other exceptions to the Act are revolvers needed by Army officers in the discharge of their duty, and pistols sold to people going abroad for a period of not less than six months. In the two latter cases, if you are not personally known to the dealer, it is legally necessary to produce a letter signed by a Justice of the Peace or police inspector.

The Pistols Act was framed to prevent the sale of cheap Belgian revolvers to schoolboys and "unauthorized persons," and although severely criticized as restricting trade, it has done an immense amount of good in preventing accidents, and should be further extended.

Latterly, in 1914, the cheap and nasty foreign pistols known as "saloon pistols" or "long-barrelled Derringers" have been made with 9½-inch barrels expressly to evade the Pistols Act. The result was almost immediate, and an "accidental death" or street-shooting case with one of these weapons occurred early in April, 1914.

As it is necessary to have a ten-shilling gun licence in order to be allowed to carry or use a firearm anywhere, except in your own "house and the curtilage thereof," the advertised sale of these weapons, for which it is advertised that "no license is necessary," deserves and needs rigorous suppression. The person who is genuinely interested in pistols can afford to purchase a licence and sound weapons, but the sale of dangerous firearms, at six or seven shillings apiece, to boys and ignorant people cannot be too strongly condemned.

These weapons are smooth-bore, ·22 calibre, single-shot pistols, made to take the No. 1 bulleted breech-cap. This little cartridge is perfectly capable of inflicting fatal injury, while its small size and mild report are anything but convincing of its genuinely dangerous nature. Such weapons are usually sold with unadjusted sights and terrific trigger-pulls, but properly adjusted they are accurate at ranges up to fifteen yards, and will kill rats, rabbits, cats, etc.

A peculiar weapon is known as the " Dedless pistol," and is made in ·410 shot-cartridge sizes, both as a single-shot and as a four-shot revolver. The alleged use of this arm is to render an assailant unconscious by means of a discharge of asphyxiating gases, and it is claimed that no fatal injury can result. The cartridge that is to accomplish this feat contains a variety of pepper, lycopodium powder, and other ingredients, and produces a loud report and enormous flash.

As the whole purpose of a firearm is to stop adversaries *at a distance*, the short-range limitations and dubious efficiency of this pseudo-weapon render it of doubtful value ; the primary assumption that the assailant must of necessity *inhale* the fumes is also uncertain of fulfilment, as the discharge might take place exactly as he finished inhaling a breath, and he probably would duck and flinch from the explosion. Its pseudo-harmless nature is an additional danger, as, fired by a fool for a " practical " joke, a wad or small foreign body in the cartridge or scales of dirt or rust in the barrel might destroy the victim's sight.

A peculiar group of pocket pistols claim attention because of their design upon lines of extraordinary thinness for the pocket. The "Reform" pistol is one of these, and consists of four superposed barrels in one block. These rise one barrel at a time when the arm is cocked, and it is worthy of note that the recoil of the succeeding shots automatically ejects the empty shells from the three top barrels without the use of mechanism, the simple shock being sufficient to throw the empties clear. The arm is chambered for the ·25 automatic cartridge.

Another queer pistol of similar design is "Ashland's Patent," a weapon of ·22 calibre, with four superposed barrels. In this weapon the barrels are fired successively by means of an ingenious series of firing pins, hit successively by means of an eccentrically-pivoted revolving hammer. The arm is non-ejector, tip-up action, and fitted with a folding trigger.

Squeezer or grip-operated repeating pistols have been known for some time, the weapon known as a "Chicago Squeezer" being the prototype. In external appearance this device resembles a large watch or tape measure. From one side projects a short barrel, and from the other a short lever, shaped to fit the palm of the hand. The cylinder is of the axial or cart-wheel type, and holds seven ·32 short rim-fire cartridges. Squeezing the pistol in the hand with the barrel projecting between the second and third fingers actuates the mechanism, so that the act of delivering a blow or punch at one's adversary, or the mere tight clenching of the fist, discharges the weapon.

A similar pistol, but with a square grip handle somewhat similar to an automatic, was popular in France. A special cartridge of approximately ·32 calibre is required for it.

Devices for surprising the highwayman or footpad are not uncommon, and pistols made to resemble pencils, pocket-books, or watches are not unknown. One of the most ingenious of these is the watch pistol, made to resemble a watch, but firing a ·25 automatic cartridge. The muzzle of the weapon is the milled head of what is apparently the winding stem. The trigger is a round bar moving in a groove, which runs back to engage the hammer. The pistol is fired by means of a simplified double action, and the projecting end of the trigger is pressed back until it loses engagement with the hammer, as is usual with the average automatic pistol-firing device. To reload, the inner tube of the barrel is removed by unscrewing the milled head of the apparent watch stem, a fresh cartridge is inserted, and the barrel screwed back into place. The weapon is a single-loader, but as a surprise for the footpad probably entirely efficient.

The disguised or concealed pistol has always been popular, and is found both in flint-lock and all conceivable ages of pistol combined with whips, swords, purses. A modern Belgian manufacture is the pin-fire revolver and sword combined.

Water-pistols barely come within the scope of

this work, but it should be remembered that when loaded with strong ammonia or capsicum tincture, they become dreadfully efficient weapons for blinding an adversary.

The sporting or shot-gun pistol is seldom made nowadays, though double-barrel under-lever pistols similar in design to shot-gun practice were very popular in South American republics as horse pistols, for use with a charge of shot or slugs from on horseback. The author uses a side-action ·410 single-shot, 10-inch barrel, shot pistol for sporting purposes, but at any range over twenty-five yards the shot scatters too widely for certain effect. The best-known modern sporting pistol is the composite weapon known as the Marbles "Game-Getter" pistol, which combines a variety of smart mechanical adjustments and a wire shoulder-stock with an under and over pair of barrels. The top barrel is smooth-bore and chambered for the ·44-·40 Winchester shot cartridge, and can be used with a special round ball; the lower is rifled to take the various lengths of ·22 rifle cartridge. The weapon is made in several barrel lengths, and is certainly ingenious; excellent results are claimed for it, but as the ·44-·40 shot-cartridge is practically an unknown charge in Europe, the arm is little known on this side of the Atlantic Ocean.

In countries where game is not shy and met at close range, and explorers can carry but little surplus weight, the single-shot pistol is sometimes carried in place of the heavier rifle. A Watson's single-shot target pistol of ·360 No. 5 calibre was used in the Hudson's Bay territory by Mr.

Alexander Crundall, and proved perfectly capable of accounting for moose and similar big game. American writers on sporting subjects are great advocates of the ·22 single-shot, on account of the extremely light weight of the ammunition; but one hesitates to depend on it for anything larger than grouse or small pot game, as to bring down larger animals a fatal head-shot is essential —a matter not always easy of accomplishment.

CHAPTER XV

LEARNING TO SHOOT

THE tyro or novice who wishes to acquire skill in pistol-shooting must remember that, unlike mathematics or any purely mental exercise, proficiency cannot be acquired by mere book learning alone. The whole basis of shooting rests upon physical efficiency, and it is a matter of training certain muscles and nerves to execute the commands issued by the brain.

This physical training of the muscles is allied with a certain amount of nerve and brain discipline, which are essential to overcome the natural or instinctive faults of incorrect eye-work and flinching from the recoil or report.

In process of time, care in following out the prescribed rules of aiming, trigger pressing, and correct position, results in these being acquired as habits, and later becoming almost instinctive. Practice makes perfect, and a fair natural shot may easily take rank as a first-class marksman after three to six months' training and practice at suitable ranges and with suitable weapons.

The great thing to avoid is acquiring bad habits during the early period of training, as once adopted these may take years to eradicate.

There are two perfectly distinct styles of shooting: the target shot style, where excessive accuracy and slow but certain shooting at bull's-eye targets is the ideal; and there is the practical pistol shot style, in which a lower criterion of marksmanship, but a very much quicker rate of fire, is the ideal. This latter style of shooting is the only one of value to the soldier, duellist, or individual concerned with self-defence; but the preliminary training for the two styles is identical, and it is unwise to attempt speed work until a fairly high standard of slow target work has been attained.

Different styles of shooting suit different individuals; and where the massive, deliberate-fire target shot may be a cold, unemotional, and phlegmatic business man, the speed shot is usually of a much more impatient and nervous temperament. The man who is strong and skilled enough to be proficient in both styles is undoubtedly the only perfect type of shot.

Having decided to take up pistol practice, you must acquire a revolver. For a novice a pocket weapon is of no use, and efficient training can only be obtained by using the long-barrelled target revolvers designed for accuracy and serious competition work. A good plan is to start with a ·22 or ·32 target pistol or revolver, as the absence of recoil and unpleasant features enable one to master the intricacies of aiming, pull-off, and the general handling of the weapon, before passing on to the standard ·455 target revolver. If, however, your purse does not warrant the purchase of a small-calibre weapon in addition to your regular competition model, commence with one of the latter.

There are only three makers to choose from—Colt, Smith and Wesson, and Webley; but the latter firm make target revolvers in standard calibre and 7½-inch barrels, with various forms of handle or grip. The whole point of selection of a target revolver is to find the one whose grip is most suited to your hand. If opportunity offers, go to a good revolver range (King's Gallery off Panton Street, Haymarket, and next to the Prince of Wales's Theatre, can be recommended), and sample various types of weapon till you find the one best suited to your individual taste.

Do not buy second-hand target weapons without expert advice, as they are probably worn-out pistols reblued to look as if in perfect condition, or old types with sights forbidden by more recent competition rules.

Having selected your pistol, the next process is to get it accurately sighted, as most target arms are sent out with fore-sights sufficiently over-size to allow for the necessary individual adjustment.

There are two types of fore-sight—the plain rectangular section blade, and the bead or under-cut type. The back-sights provided with the pistol consist of a plain bar, a **U**-notched bar, or a square **U**-notch for use with the square fore-sight. Higher sights are supplied for longer ranges. Having ascertained by practical trial which form of fore-sight suits you best, stick always to that type, and never be tempted to chop and change about; but be sure that it is of a type allowed by the N.R.A. rules. A beginner must of necessity have his sights altered for him by an expert; but later on, when reliance can be placed on his own shooting, all changes should be determined upon from his own personal practice with the weapon.

The majority of shooting takes place upon the 20-yard range, and it is for this distance that adjustment should be effected.

The great thing to remember is that the front-sight should stand out clear in the back-sight and against the target, and that you should *see the same amount of it* each time. If at one shot you see one height and at the next less, the shots will not maintain the same elevation. This vertical adjustment is the main and most important point.

A revolver is tested by firing a group of shots at the aiming-point—that is to say, with the top of the fore-sight clear in the notch of the back-sight, and aligned upon the bottom edge of the bull. The direction of shots upon the target is indicated as is the face of a watch; so the bottom of the centre of the card is six o'clock, the centre of the left edge nine o'clock, and so on. The correct point of aim is six o'clock, at the bottom of the bull, and the bullet should strike one inch higher—that is to say, in the centre of the bull.

Supposing that the centre of your group of shots on the card is 2 inches too low, and 2 inches to the left of the centre of the bull, it is obvious that you must adjust your pistol to shoot 2 inches higher and 2 inches to the right.

Vertical adjustment is secured by filing the fore-sight, lateral adjustment by tapping the sliding back-sight into the desired line.

In altering sights, remember the following points: *Fore-sight follows back-sight*—that is to say, deepening or lowering the **U** in the back-sight makes you shoot lower, raising it makes you shoot higher.

Filing the fore-sight down makes you shoot *higher*. A higher fore-sight makes you shoot *lower*.

The formula to determine the amount to alter is: As range is to sight radius, so is the error on the target to the height of fore-sight, all dimensions being in the same denomination.

As an example of this, if the sight radius (length from back-sight to fore-sight) is 6 inches and the range 20 yards (720 inches), and your group of shots is 3 inches below the bull on the target, then the simple proportion sums $\frac{6 \text{ inches} \times 3 \text{ inches}}{720 \text{ inches}} = \frac{1}{40}$ inch gives the amount—$\frac{1}{40}$ inch—that must be filed off the fore-sight.

Final adjustments should be done with a file rather than by formula; and in widening the **U** remember that cutting metal away on one side makes the weapon shoot toward that side. Deep near-sights are undesirable, and the **U** should be little more than a semicircle; if a square or rectangular notch is used, the depth should be proportionate to the width. In all cases the front-sight must stand clear of both sides of back notch, as otherwise you cannot see if you are not covering more of the front-sight on one side or the other—a fault which will cause errors in lateral direction.

The bead-sight is sometimes called the "Ira Paine" fore-sight, and it is amusing to note that the name of an American has been recently applied to the ordinary square-notch and rectangular fore-sight, although this pattern was popular and common on muzzle-loading weapons a hundred years ago.

The Winans fore-sight is another ingenious device, which effects the same purpose of seeing always the same amount of fore-sight, as the undercut bead. A short portion of the upper part of the face of the fore-sight is cut vertically, or recessed, so as to cause it to appear black, as if in shadow; the remainder of the face is curved, so as to reflect the light. The standard military or service revolver sight can, by filing, be readily converted to the Winans pattern, without any liability to catch in the clothing or holster.

Target sights should have as sharp and severe an edge as possible, and the edge of the **U** or notch furthest from the eye should be bevelled or chamfered, in order to prevent any tendency to blur.

The only other point likely to require attention is the trigger-pull. This is usually excessive when sent out from the makers, and should be perfectly sweet and easy, with no trace of "drag" or "creep." The regulation Bisley pull is 4 pounds, and it is well to have the adjustment made to an ounce or two heavier, in order to avoid being disbarred should wear slightly lower the pull-off till it is below the official standard.

For target pistols, not for competition use, a 2½ pound pull is excellent and sweet to handle, but the lighter the pull-off the more dangerous the arm is in unskilled hands. A target-shot should always accustom himself to the official standard, as, if accustomed to a light pull, it will affect his shooting adversely to return to a heavier one.

Amateurs should never attempt to adjust their own pull-off, as it needs a skilled gunsmith to perform the delicate operation. All tinkering or investigation of the mechanism should be sternly prohibited, and the weapon treated with as much care and respect as if it was a valuable watch.

The correct conduct of revolver shooting is as important a matter of etiquette as correct table manners, and a foolish man handling weapons in such a manner as to endanger himself or other folk, either out of ignorance or foolhardiness, is a pitiable sight. It is not uncommon for boastful idiots to enter a public range and attempt to impress their audience by shooting with a variety of stage flourishes, swinging the revolver muzzle back over the right shoulder in such a manner that a spectator behind might easily be shot. Such a man should be instantly called to order, and, if he continues the practice, asked to leave the gallery. One may suffer fools gladly in many rôles, but not when they have loaded firearms in their hands.

A revolver should always be treated as if loaded, never left loaded, and always held with the barrel pointing toward the target.

On picking up a revolver, whether your own or a strange weapon, *always* swing out the cylinder in order to see if it is loaded. It does not matter what you *know* or *think*—make *sure.*

The danger of the revolver as regards accident is greater than that of any other firearm, except the automatic pistol, which has the same shortcomings, with the added danger of an overlooked cartridge in magazine or chamber. All revolvers and pistols are more dangerous than guns or rifles because of the short barrel length, which enables it to be turned in a very close radius without elevating the barrel, as with a rifle.

The weapon should never, in any circumstance, be left with the barrel pointing anywhere but down the range; it should be loaded in this position and fired in this position, and if lowered to rest on the ledge or counter of the range between shots, held in this position. Sometimes men will cock a revolver with both hands, and in doing so will swing round with the barrel pointing direct at their neighbour. No more dangerous trick can be performed, and double-handed cocking is the hall-mark of ignorance and inefficiency. Learn to cock a pistol by simple thumb pressure of the holding hand, the weapon resting meanwhile upon the ledge in front of you, with its barrel pointing toward the target. Double-action work, or firing by means of the trigger-pull, cocking and discharging the arm, is only to be practised later on. All preliminary target practice is performed, using the weapon as a single-action arm, and cocking must be learnt. At first it may seem difficult, but it is nearly all "knack," and a very little practice will develop the thumb and forefinger muscles till the practice becomes perfectly natural and easy.

Practise cocking with an empty revolver until you are familiar with the action ; also go through the usual motions of breaking the action or swinging out the cylinder and ejecting the cartridges till you are familiar with the arm and can do all the requisite motions, *without pointing the barrel anywhere but down the range toward the target.* Always cock the pistol steadily, not with a jerk, as if the hammer slips ; it is likely to injure the bent and scear. When a pistol is properly cocked, the hammer should come back slightly beyond full-cock, when a click is heard, due to the bent falling into the hammer notch. The hammer should then be released, and will remain in position. During cocking the first finger should be kept well clear of the trigger.

Having mastered these preliminaries, load your weapon, take your position at the range—preferably for beginners at the short or 12-yard target—at which the standard 20-yard card with 2-inch bull is used. Short ranges are best to begin with, as encouraging success is easier of attainment.

The position, or stance, taken up is a matter of great importance, as it is essential to cultivate an easy, natural position, which is yet the best for shooting purposes. The old duelling or swordsman's position, with the right side held toward the target and the head turned at right angles to its natural position, is now quite abandoned for anything except wax-bullet duelling practice, when the narrower the target presented to an adversary's pistol the better.

The best position is to stand slightly to the left of the target, with the right shoulder slightly advanced, and the body at an angle of possibly 30 degrees. The right foot should be forward, and in line with the target; the left about 7 or 8 inches to the rear, and almost square to the other foot.

The left hand may either rest upon the left hip or the whole arm hang loose, the hand behind the left buttock. The position should be solid and natural, the weight of the trunk supported evenly by both legs, and the right arm either fully extended or, as is more usual, slightly flexed.

The grip, or correct way of holding a revolver or pistol, is all-important. The best grip is that which brings your hand as much as possible in line with the axis of the barrel. The thumb should be extended as far as possible, *and as high up as possible*, along the left side of the lock, and the trigger pressed (not pulled) by the first finger.

Different arms have different values for varying hands. In some hands the W.G. Target trigger-guard is too near the stock for comfort. Some people find the Colt handle too big for comfort, others complain of the cutting recoil of the S. and W. on the web between thumb and first finger; but the golden rule is to adopt the weapon that suits you best, and develop a grip to suit it. In the Colt, a favourite position is with the little finger curled up beneath the butt, a position of grip that needs considerable practice before it becomes comfortable. I, personally, prerer the W.G. Target to the Webley Scott, with the new model handle; but the question of grip is one of individual hands rather than individual designs of pistol, for it is obvious that a slender stocked pistol, suitable to a slender-fingered, long-thumbed, artistic hand, may not suit a hand like a leg-of-mutton and fingers like carrots, as well as the big and clumsy grip, of another weapon. The fully extended thumb position is only adaptable to revolvers where the action-lever or catch is out of the way, and in most modern weapons a slightly crooked thumb position is essential.

Take aim at the bottom of the bull, and, holding the revolver upright and *perfectly level*, take a deep breath, brace all the muscles of your arm and right shoulder, and *slowly increase the pressure of the first finger on the trigger, squeezing straight back, not sideways.*

The top of the front-sight, resting in the notch of the back-sight, with an equal margin of daylight on each side, should just touch the bottom of the bull; and if the pistol is properly sighted and held, and the trigger properly pressed, the shot should find the bull. Probably you will find that you wobbled all over the place, and, when you did fire, the front-sight jerked off the bull with the fall of the hammer, and the shot went wide. The recoil of the weapon may have left your hand in a shaky state, but even so continue practising for a few more shots.

Always remember that each shot should be carefully and studiously aimed, and that all effort must centre upon getting a steady pull-off. The

German Army instruct recruits to stiffen all the muscles of the arm and all the fingers to a tight grip on the stock, the pressure being gradually transmitted to the trigger finger in the form of a slow and gradual *squeeze of the whole hand.* The usual cause of bad shooting is "pulling-off," and it can only be overcome by continual practice.

Never allow your nerves to "flinch" from the anticipated recoil; this is the cause of much bad "pulling-off" and bad shooting. Custom will soon get you used to the explosion of recoil, and, if you are physically fit, nervous tension should not be enough to cause flinching when you are accustomed to your weapon.

Never attempt to improve your shooting by varying your aim, by taking different amounts of sight, or aiming at a different point on the target. Centre all your abilities upon getting a good group—that is to say, as close together in one place as possible. When you can reasonably accomplish a group not larger than an orange, then you may begin to practise applying this group to the bull. Get a skilled shot to try your revolver for you, as it may not be correctly sighted; a touch or two with the file and an alteration of the back-sight should soon accomplish this, as it is usually the fault of the revolver, not the shooter, if the group is not in the right place, once a constant group can be made.

As time goes on your groups will get smaller and smaller, and you will begin to do better work at the longer range target, until you develop into a good average shot. If you think that you have reached the limit of good shooting possible with your physique, a course of grip dumb-bells or carrying a steel walking-stick may vastly improve the tone of your muscular system. Good shooting and poor health never go together. A good shot must be moderately free from bad habits, as drinking, smoking, etc., in excess affect both nerves and eyesight; but the best training for shooting is continuous and regular walking exercise. Such relaxations as driving motor-cars or riding motor-cycles leave the arm muscles and nerves too jumpy and tired for good shooting to be done.

Open-air shooting is more difficult than gallery work, as windage, varying lights, etc., double the difficulties. It is as well to practise on an open-air range whenever possible, and a seabeach, or a quarry, or chalk-pit, make good outdoor ranges.

Shooting in enclosed country at a target pinned on a post or tree is exceedingly dangerous, as bullets may ricochet or glance off and wound people a long distance off.

Casual practice of this kind cannot be too deeply deplored, and all good revolver shots should be careful to practise only in the seclusion and safety of a proper range, chalk-pit, quarry, or similar safe place.

CHAPTER XVI

SPEED SHOOTING

THERE is such a wide difference between speed shooting and target shooting, that it is best to treat the former section under a separate chapter heading. Of all *practical* pistol shooting, speed work is the only section worth considering, for where your deliberate, hair-trigger target shot may be capable of immense accuracy, he is seldom a quick or instinctively accurate shot at a moving target that traverses his line of aim.

Shots at a target that bobs up and down in the same place are fairly easy; but for practical work with the revolver as a weapon of offence or defence and for sporting purposes, it is necessary that actual aiming should be almost done away with, and the whole training devoted to developing an instinctive facility with your weapon, which should enable you to put all six shots on a 12-inch card at 20 yards within three or four seconds.

The pistol is pre-eminently a weapon designed for quick use at close quarters, and the "target-shooting" side, which has developed the present excellent modern revolvers and pistols, is certainly an important branch of the art of pistol shooting, but by no means the most practical.

If you had to choose between entering into a revolver duel at ten or twelve paces, with an expert target shot accustomed to take possibly thirty seconds or more of time to take proper aim, dwell on the target, and eventually press trigger; and meeting a man who was not a target shot, but accustomed to use or carry a revolver for self-defence work in wild countries, it would be better to select the target shot as an adversary. The self-defence man would be fairly certain of not taking more than thirty seconds to discharge the whole six shots, and place them within the radius of a 4-inch circle.

The old Western practice of throwing up a tomato-tin into the air and emptying one's pistol at it, so as to keep it continually moving along the ground after it had fallen, was a very useful form of training, as it taught one to shoot instinctively without so much as looking at the sights, depending on the balance and set of the weapon in one's hand to enable one to shoot and hit instinctively and without mechanical aim, exactly as one throws a stone or cricket-ball.

When you reflect that an expert can fire the six shots out of a Webley Target revolver and get them all on a 12-inch square card at 12 yards *in less than two seconds*, and that the eight shots of an automatic pistol can be equally accurately discharged in less than *one second*, the extraordinary effectiveness of the revolver or pistol as a weapon when handled by an expert can be readily realized.

The average man refuses to believe that such speeds are humanly possible, and reinforces his argument by the claim that the double action of a revolver cannot possibly be functioned by the human finger at this speed. He actually means that the muscles of the finger and eye cannot obey the brain commands necessary for each separate discharge of the six shots. In this he is technically right; but the actual process that makes the feats of speed work possible is—that the eye receives only one command-*look straight at the target*, and does not need a separate "brain message" for each shot. Next, the trigger finger of an expert operates the mechanism of a revolver much more easily and instinctively than can that of a novice, the muscles of the trigger finger being perfectly trained to their work by previous practice. Then there is the rather complex process of utilizing the recoil of a revolver, translated through the various arm muscles to the trigger finger to operate the double action almost to the point of discharge, before the barrel is level with the target for the next shot.

Just as Press photographers develop a knack of instantaneously pressing the button of their reflex cameras *a fraction of a second before* whatever they desire to snap happens, so the speed shot *anticipates* the falling in line of target and weapon, and presses his trigger at exactly the right instant.

To fire the six shots at different targets takes infinitely longer than to fire the six at one, the time being lost by the transmission of the brain orders to change direction to the muscles as well as by the actual traverse of the weapon.

The recoil of a shot naturally throws one off one's aim, but by practice it can be absorbed, just as one absorbs the shock of a caught cricket-ball, and the muscles trained to bring the weapon automatically back into correct alignment with the target. This is the first point to remember—always accustom yourself to bring the weapon back to the point of aim after each discharge.

Next master the practice of double-action shooting. In this a firm grip of the weapon is taken and steadily increased as the trigger is squeezed back. You will soon get to "feel" intuitively the end of the travel and the imminence of the fall of the hammer. This "feel" of the exact moment previous to the fall of the hammer varies with different weapons, as some have shorter or longer hammer falls than others. In some pistols, notably the ·38 hammerless Smith and Wesson, a distinct pause can be felt after the hammer has reached the top of its stroke, and before the continued trigger pressure pulls the scear clear of its bent.

Speed practice should be commenced at the short range, and continually varied by practice at two or three targets arranged unevenly at the same distance. A good practice is to let the marker give the command at which target the shot is to be delivered, as the varying order of the commands—Left, Right, Centre, as his fancy dictates—insure more practical shooting than plain high-speed discharge at the same target.

Always work against a stop-watch, and vary your practice, firing some targets for speed alone and some for speed and accuracy combined. The various so-called "quick shoots" at Bisley are archaic, as they are really pure target competitions, with a limit of time in which to discharge the shots. Two and three seconds per shot are frequently allowed, and there is no scale of proportion between time and accuracy.

The "bull's-eye" revolver competition is of little practical value, and the substitution of speed competitions fired on a man-sized target at 20 and 40 yards ranges would have many times the actual military value of the present competitions.

In France and the U.S.A. the proper military value of quick shooting and man-sized targets is recognized, and it is to be hoped that the new military programme of the control of the N.R.A. Bisley meetings will evolve an efficient series of reforms, and that a proper standard course of revolver instruction and practice at proper targets will be adopted by the military authorities.

The U.S.A. revolver course is fired at "Target D," a silhouette figure of a standing man. Another target, "Target K," the silhouette of a mounted man, is also used. These targets are set up in groups at varying distances and varying angles to the line of movement of the shooter, who has to ride past them at the trot, walk, and gallop, firing one shot at each as he passes.

In the U.S.A. Navy practice is carried out at "Target A," a 6-foot by 4-foot rectangle, with an 8-inch black bull's-eye, counting 5, and other concentric circles of count : 26-inch ring counts 4, 46-inch ring 3, the rest of target 3.

The sharpshooter's course, both for instruction practices and annual record, has to be fired on this target—6 shots at 15, 25, and 50 yards. *Time limit*, 18 *seconds for* 6 *shots*.

The U.S.A. National Guard, which corresponds to our Territorial organization, uses the same target, and grades its men into three classes: Marksman, sharpshooter, and expert. To qualify as marksman or sharpshooter, 65 and 80 per cent. of the possible score of 200 points must be made respectively. The course is as follows:

15 yards,	two	scores	of five	shots,	10	seconds	per five shots.
25	”	”	”	”	10	”	” ”
25	”	”	”	”	30	”	” ”
50	”	”	”	”	slow fire 1 minute per shot.		

The expert qualification is only open to those who have passed the previous test as sharpshooters:

15 yards,	two	scores	of five	shots,	8	seconds	per five shots.
25	”	”	”	”	8	”	” ”
25	”	”	”	”	20	”	” ”
50	”	”	”	”	20	”	” ”
75	”	”	”	”	20	”	to each shot.

An expert must make 80 per cent. of the possible score—200 out of 250 points.

The French military practice is carried out both at silhouette and ring targets, but at Gastinne Renettes all duelling pistol practice is performed at special electrically-indicating silhouette targets. Practice is performed on the "word of command" system, as in the recognized code laid down for duelling, described later in the section on wax-bullet duelling practice.

The silhouette targets have varying values according to the portion struck, and the usual range is 25 paces (19 metres).

The well-known wax-bullet duelling practice is the outcome of research and experiment on the part of Monsieur le Docteur Devillers, whose inventions have rendered it possible to obtain actual duelling practice and exchange shots with an adversary without actual harm resulting. The system employed consists of fitting breech-loading duelling pistols of ·44 calibre with a special steel cartridge bush—or adapter. The fore part of this is hollowed out to accommodate a bullet made out of a special soft wax that has been evolved after much experiment, and experiments in the use of other materials, such as papier-maché, are still proceeding. The difficulty was to find a material that would not shatter from the force of the explosion, would not deform in the barrel, and was yet soft enough not to bruise the person at whom it was discharged.

In 1904 M. de Docteur Paul Devillers had discovered the necessary materials and founded the Society of "L'Assaut au Pistolet" for the practice of wax-bullet duelling and pistol shooting in general.

The charge employed is of necessity very light, and consists of a special ·22 calibre cap, which is loaded into the base of the steel cartridge.

In order to make this cap centre-fire, two projections of the shell are doubled back inside it in such a manner as to form an anvil. The wax bullet is pressed lightly in a cup on the opposite side of the adapter and the whole combination inserted in the pistol. These bushes or adapters are made in a variety of calibres to fit various revolvers and pistols, but the practice is carried out with a specially designed pistol that is modelled as closely as possible upon the standard muzzle-loading duelling pistol.

These and all wax-bullet fittings and ammunition are supplied by Piot-Lepage, 12, Rue Martel, Paris, the official armourer of the Society.

The weapon used is a side-lever-action, breech-loading, duelling pistol of ·44 calibre, fitted with a special steel-plate heart-shaped guard that completely shields the firer's hand, but does not obscure the free view of the sights and barrel. This guard is fitted just in front of the trigger-guard, and is essential, as the wax bullet, though innocuous when its flight is arrested by soft drapery, is yet quite powerful enough to bruise or break a finger clenched against a hard substance, such as a pistol stock, which would act as an anvil.

The rest of the outfit consists of a fencing-mask fitted with a thick glass window and a loose overcoat or overall of thin but loosely draped material.

The process of practice is regulated by a special code of rules, and all the special observances of correct duelling etiquette are observed.

The order of the turns in which combatants shall shoot, pool shoots, and all the pairing and handicapping of members and general arrangements, are under the charge of *le directeur du combat*, who is assisted by two *commissaires de tir*, who take the place of seconds, and a third member, who is the official loader.

The two members about to exchange shots take up their positions cloaked and masked at twenty-five paces from one another. The two seconds are charged with the duties of handing their man his weapon and observing the effect of his shot. The director of combat gives the order to fire, and decides which pistol was the first to be fired.

Having received his pistol from his second, the duellist cocks it, and takes up his position or stance. This is given in the rules, as the feet should be 70 centimetres or more apart, measured from heel to heel. The pistol arm should hang down beside the body, with the barrel pointing to the ground, the butt touching the thigh.

Under no condition may the pull-off or loading of a weapon be examined or tried by the duellist, nor may the weapon be moved from its position, with the butt upon the thigh, before the command, "Fire."

The director asks the duellists:

"Are you ready?"

To which they, having cocked their weapons, reply:

"Yes." Both must answer.

The director then gives the word:

"Fire! One!—Two!—Three!" and shots must be exchanged between the first sound of the word "fire" and the last of the word "three."

The time for the firing period is judged by the director from a metronome, and the complete sentence, "Fire! One!—Two!—Three!" is calculated to occupy four beats of the instrument. The lowest rate of command is to a metronome rate, equivalent to sixty beats to a minute; the average, one hundred; and the quickest possible, one hundred and twenty.

Thus, the highest rate of fire allows two seconds in which to raise the arm, aim, and shoot.

The period of time between the last duellists' answer of "Yes" to the "Are you ready?" question and the command "Fire," should not exceed five seconds.

This wax-bullet practice is by no means as easy as it seems. The whole process of "fire at the word of command," and the fact that your opponent also shoots back at you, tends to unsteady one's nerves.

After a bit one gets used to it, and with practice satisfactorily "fatal" shooting can be effected by merely butting the elbow upon the hip-joint, and quickly swinging up the pistol and pressing the trigger. Nearly all the members adopt a crouching, feather-edge position, with the head bent low. This is in order to offer as little target space as possible to one's adversary.

When this wax-bullet duelling with revolvers in place of single-shot pistols is practised, the seconds watch the fall and rise of the hammers of their principals, stop-watch in hand, timing from the first to last shot fired within the time of command. An additional second to each principal counts the number of shots that hit.

The score is then worked out upon a scale ot ime in fifths of a second as against hits. The score is proportional to the number of hits, and inversely proportional to the time. Thus, a man making six hits in six seconds scores 400, and a man making five hits in three seconds 667. For details, see the scale in the Appendix.

For the man who wishes to become a competent pistol shot, and not a specialized competition expert, quick or speed shooting is the best system to follow, and certainly the only one suitable for modern automatic pistols—a weapon built especially for speed work. In any case, the preliminary training must be the same, but for really practical use of a weapon, a skilled speed shot is a more competent man than the purely target exponent.

CHAPTER XVII

DUELLING

THE practice of duelling with pistols developed out of the mixed type of mounted duel when the two combatants fought with all arms on horseback, as was customary in cavalry actions of the period. Such a combat was fought between Colonel Jonah Barrington and Mr. Gilbert in Ireland in 1759. The combat took place on horseback, the weapons being two horse-pistols loaded with ball and swan-drops, swords of the contemporary cutting type, and "skeenes"—long broad-bladed daggers. Colonel Barrington was shot in the face at the first encounter, but the large ball missed him, and the swan-drops merely broke the skin. He was ultimately victorious.

The latter half of the eighteenth century and the first of the nineteenth were notorious for duels, Ireland being particularly ill-famed in this respect. At the period of 1870 no young Irish gentleman was considered a man until he had been "out" and "blazed" at least on one occasion. Duelling pistols were family heirlooms, and had special pet names, such as "Sweet Lips" and "Darling." In 1777 the Irish "Code Duello," upon which all English codes were founded, was formulated at the Clonmel Assizes, and adopted as the standard regulating affairs of honour. The usual pistol was a 9- or 10-inch barrelled smooth-bore weapon pistol, flint, and (later) percussion. Rifled pistols were regarded as bad form, and were barred by convention; but "semi-rifled"—that is to say, pistols with the breech-end of the barrel rifled to about half-way up—were occasionally made and employed.

Duelling was popular with all classes—gentry and professional. In the Army it was so rife that it was partly checked in 1792, and forbidden on pain of cashiering in 1842. Among notable pistol duels, one of the first is that between Lady de Nestle and the Countess of Polignac in 1721. Later affairs of note were John Wilkes and Lord Talbot in 1761; Earl Bellamont and Lord Townsend at Marylebone in 1773. Duels in the dark were not at all unusual, one of the most notable meetings being that of Henry Gratton and Isaac Cory, both members of the Irish Parliament.

The clergy were not exempt from the practice, and in 1764 the Rev. Thomas Hill was called out by Cornet Gardner, of the Carabineers, for "ungentlemanly conduct," and killed. The Rev. Henry Bate, of Ely Cathedral, who died in 1824, fought many duels (mostly about ladies and actresses), and killed three men.

Journalists were exceedingly liable to be called out; and even to this day duels between Continental editors and offended public characters are frequent, though often not severe. In 1821 Mr. Scott, editor of the *London Magazine*, was fatally wounded in a duel in the dark at Chalk Farm, his opponent being a barrister named Christie; and in 1835 Mr. Black, of the *Morning Chronicle*, and Roebuck, M.P., exchanged shots.

In the United States duelling was equally rife, but the "Code Duello" was slackly administered, most unusual and diverse weapons being used. A wide and disreputable latitude of conduct was permitted and condoned by popular opinion, so that the margin between homicide and duel was exceedingly confused.

During the Tripoli affair and at Gibraltar in 1819 duelling between American and English officers was common; and after Waterloo, the daily spitting, or *embrochement*, in duels of our young Englishmen by the enraged French was a scandal, only put down by the British insisting upon pistols, a weapon in whose use they excelled the French as much as the latter excelled them with the smallsword. As the French were the aggressors, and choice of arms devolved upon the challenged, this arrangement was quite fair.

The three most important duels in the States were those between Alexander Hamilton and Aaron Burr in 1804, Stephen Decatur and James Burrows in 1820, and D. C. Broderick and David S. Terry in 1859. The political issues affected by the fatal results of these duels in every case reflected upon the history of the nation.

The celebrated Bowie duel, which gave the name to the Bowie knife, was a mixed pistol, knife, and sword-stick affair, with several participants, and occurred in August, 1827.

American editors had to be good shots; and a famous affair took place between R. F. Beine, of the *Richmond State*, and W. C. Elam, of the *Wing*, as late as 1883.

The short-barrelled Derringer was a favourite weapon, and was used at short range over a table between the combatants. Such a duel was fought between a doctor and General Bankhead Magruder in 1852. American ladies also had their affairs, and a duel between two of them at Buffalo, New York, both using "Allen's Revolvers," was stopped by the arrest of the combatants.

The decline of duelling set in about 1840; but in a book by Major Ben. C. Truman, of 1884, we read: "At the present time duelling is at a great discount in Ireland, and the laws against the custom rigidly enforced."

Andrew Steinmetz, who wrote in 1868 a large treatise on the subject, deals with duelling as a modern affair, and one still practised. There was a curious terminology applied: "Leeching" was the duelling term for taking up one's firing position. To "delope" was to fire in the air; while the meeting-place was termed the "re-leager." Gentlemen perfecting their education or likely to have an affair of honour went through a severe course of training. Practice was made at small white wafers about two inches in diameter pasted on a black background, and the novice was advised to "culp," or break, three dozen of these each morning before breakfast. When he could be sure of breaking twelve of these at fourteen yards in six minutes, he himself reloading his pistol between each discharge, he could consider himself trained.

A duellist about to fight was enjoined to retire

early to bed, be called at five and take a strong cup of coffee and a biscuit, to wash the eyes in cold water, and not to wear flannel underclothes. This latter because of the danger of flannel entering the wound with the ball. To steady his nerves he may smoke a cigar, and if faint take a small brandy with effervescent soda-water on his way to the releager.

The actual duelling position varied, as there were several systems. A "volonté à marche interromper," "à ligne parallère," being systems in which the combatants advanced towards one another, exchanging shots according to a pre-arranged code.

"Au signal" was the usual straightforward method of firing on the word of command, and "à barrière" the equally popular method of placing the combatants back to back, whence at a word they marched ten paces forward, turned round and fired.

In the usual "au signal" duel the adversaries were placed facing one another, and either both fired at the word of command, or it was decided by a tossed coin or the Code Duello which fired first. In no case were they allowed to take dead aim at one another before the word to fire was given, but each had to hold his pistol pointing to the ground behind his back, or erect armed, pointing to the sky. Ten, twelve, or fourteen paces were the usual distances.

The duel "à l'outrance" was the only exception to the above rules. In this the combatants each held one end of a handkerchief, and presented weapons at each other's breasts before the word was given. Only one pistol was loaded, selection being made by lot; the winner having the choice of taking whichever weapon he pleased, both being covered by a handkerchief.

The duelling of the old days was covered by a strict code, but it was customary to bring your own pistols, whereas, nowadays, new pistols unknown to the combatants are insisted upon. Possibly the best accounts of duels are those given in fiction by authors who lived during the duelling period. A man of that time writing a book would of necessity be accurate in his description of the details and mode of procedure.

Samuel Lover in "Handy Andy," 1842, gives an account of Irish contemporary life during the period 1820-1840, and the duels therein are fought with flint-locks; the combatants bringing their own weapons to the ground, the seconds loading for them, and both combatants "blazing" at the word of command.

The custom of duelling is now happily almost extinct, for it was a bad custom, and did more harm than good. It is often urged that there are cases when duelling offers the only decent way to settle an affair or punish a gross offence. There is much to be said for this point of view, but it was this same view-point, carried to impossible extremes, that forced public opinion into legislation which classes duelling with murder.

If one could trust that the Code Duello would only be appealed to in certain extreme cases, it would be an excellent antiseptic against certain conditions of our social development. Nowadays one is no longer obliged to hazard one's life "in honour bound" over some trivial matter, but during the duelling period one was at the mercy of fools and perforce had to be shot at over trivialities.

To carry a challenge is now a severe offence against this country's laws—to accept one even graver. Participation in an affair of honour would insure any officer's dismissal from His Majesty's Service, and to all intents duelling among British folk is obsolete.

Duels are still permissible on the Continent, and are not quite the bloodless affairs they are thought to be by the public.

Sword duels with the "épee de combat" are not as a rule serious, each opponent trying to pink the other in the forearm. Duels with sabres are much more serious, and duels with pistols the most serious of all.

In good class European circles it is now a rule that duels must be conducted as privately and with as little self-advertisement as possible, and the affairs are usually regulated by a Court of Honour that does its best to bring about a reconciliation between the parties.

Assuming that an honourable settlement is impossible, the seconds, two friends to each combatant, arrange the place and time of meeting. A case of new regulation pistols are bought at a suitable gunmaker's, loaded in the shop, and the case sealed to prevent tampering with the load or practice with the weapons.

The usual distance is 25 metres, and choice of position for their men is settled among the seconds by drawing lots, as one place may have an advantage in light or wind.

The duel is conducted by a gentleman who is not a second : " M. le Directeur du Combat." He opens the pistol-case, and withdrawing the weapons, offers the choice of the pair to the combatant who has drawn the winning lot. The other weapon is handed to his opponent.

They take their places, and at command cock their pistols.

The director stands back from the combatants about midway, and gives the command just as in the wax-bullet practice :

" Attention !—Feu ! Un !—Deux !—Trois !"

The combatants must raise their pistols and fire *after* the word " feu," and *before* the word " trois " is finished.

The Court of Honour, who regulate the affair, often see that the words of command are given very quickly, in order to disconcert and hurry the duellists as much as possible. The normal rate is at a hundred words to the minute, but it can be said infinitely faster.

If both miss, and one or other desires a second shot, they change places ; the weapons are reloaded, and further shots exchanged.

Under this system there is little time for " deloping "—*i.e.*, firing in the air, but the same condition can be intimated by placing one's pistol behind one's back on the word " feu."

Modern duels in which one or both of the combatants are English do occasionally take place, and are always fought on foreign soil. To avoid all legal risk the challenge and its acceptance should also take place abroad, for it is not enough that the duel alone is fought abroad if the affair was started on British soil.

In any case the duel must be carried out under the existing rules of the country in which it is fought; and the use of other weapons or any informality of procedure would render all parties liable to arrest and punishment.

CHAPTER XVIII

EQUIPMENT, ACCESSORIES, CLEANING, ETC.

THE first secret of good shooting is the possession of a first-class weapon, and unless great care and attention is paid to the cleaning of firearms they rapidly degenerate in efficiency.

Revolvers and pistols must be kept entirely free from rust and leading, and to insure this it is essential that the weapon should be cleaned as soon as you have finished shooting.

In the old black powder days the amount of fouling produced by each discharge was enormous, and left a black residue of unburnt carbon and potassium sulphides in the barrel.

Each succeeding shot increased the thickness of the deposit of fouling, till after twelve shots the accuracy of the weapon was perceptibly affected and the recoil considerably increased.

Black-powder fouling was hard to remove, except by the use of boiling water. In the days of muzzle-loading single-shot pistols it was easy to detach the barrel and breech-block from the stock and plunge the whole in boiling water, which speedily scoured out the soluble deposit. The barrel was then withdrawn by means of a wooden stick pushed into the bore and set on end to drain and dry. The hot metal speedily evaporated the remaining water, and left the barrel clean and dry. It was then oiled over inside and out and remounted on the stock.

With the increasedly complicated mechanism of revolvers, the danger of getting water into the lock or ejecting mechanism, where it would be overlooked and likely to rust the delicate parts, led to its disuse, and black-powder revolvers were cleaned with tow and oil or bristle brushes in the ordinary dry way. This meant a good deal of messiness, and was a tiresome job.

Nowadays, the fouling left by smokeless powder is so little that round after round can be fired without its affecting the accuracy of the weapon; but the fouling of all smokeless powders has particular vices of its own.

The black-powder fouling was not instantly injurious to the barrel metal, as it consisted for the most part of neutral or alkaline salts. The residue of smokeless is, on the other hand, acid, and immediately attacks the metal of the barrel and chambers.

Smokeless powder is composed of varying proportions of nitro-cellulose and nitro-glycerin, dissolved together with acetone, and pressed into various forms, such as grains, sticks, discs, etc. These, on explosion, liberate a quantity of nitric acid gases, which permeate the residue or fouling, and are exceptionally corrosive.

To get rid of these it is necessary to clean one's weapons with "nitro-solvent" or "cordite cleanser," which are merely blends of mineral oil, some alkali, such as caustic potash and phenol or carbolic compounds. These alkalis render the acid salts of the residue inert and easily removable.

All ·22 rim-fire cartridges exercise great corrosive effects, as the heavy charge of fulminate of mercury liberates a variety of acid gases, and cordite cleansers should always be used to clean weapons used with this cartridge.

A cleaning kit for each calibre of weapon that you use is the best equipment, but on occasion a single cleaning rod to fit them all, and special bristle-brushes for each, will serve.

The requisites are a twisted wire bristle-brush fitting the bore fairly tightly, a cleaning rod of brass or soft metal, or sometimes wood for large bores, to which screws a jag for patches and a loop for holding a special oil patch, a bottle of special gun oil, and one of cordite cleanser.

The great point is to avoid all chance of damage to the edges of the rifling, and it is always best to clean weapons with break actions from the breech end, as the slightest damage to the rifling, near the muzzle, may affect the shooting badly.

After shooting, clean out the barrel and chambers with a bristle-brush dipped in cordite cleanser. Then clean with flannel patches wrapped round the jag of the cleaning rod, until they come through the barrel clean, showing no signs of fouling or dirt of any kind. It should be possible to draw a white silk handkerchief through the chambers and barrel of a properly cleaned revolver, without it showing the slightest speck of dirt. When it is clean, and before putting it away, oil it thoroughly inside and out.

Pistols, even if unused, should be examined and cleaned once again about three days after putting them away, in case a little fouling has been overlooked, and even if kept in air-tight cases an examination every month is a good rule to go by.

Apart from fouling, there is a disagreeable feature known as "leading," which is particularly noticeable in small-bore high-velocity arms, such as automatic pistols that fire a nickel patched bullet, and to a greater or lesser degree in all revolvers and pistols.

This leading consists of the surface of the rifling becoming covered with a thin plating of lead or nickel, which is hard to see when you are cleaning, as it takes the same metallic polish as the barrel itself. Underneath this plating there is always a certain amount of acid fouling which eats away the barrel, causing slight "pitting" or corrosion marks, which will soon ruin the barrel for good shooting.

To avoid this, careful cleaning is usually enough, but should it occur the only thing to do is to clean the barrel with any of the special leading or "cupro-nickel solvents" sold for the purpose. In ·22 weapons, where the leading occurs frequently, cleaning occasionally with a mercury paste, or simply filling the barrel up over-night with mercury, will easily remove it, as the quicksilver combines to form an amalgam with the metallic lead, and does not affect the steel barrel.

When cleaning a weapon do not forget to clean the breech of the barrel and the mouths of the chambers, as well as the inside of barrel and chambers. At these points wear occurs quickly, due to the escape of gas between barrel and cylinder, and neglect increases the size of the gap and decreases the shooting capacities of the weapon.

Pistols should always be kept in wood or leather cases, not left loose in their holsters. These cases should be as nearly air-tight as possible, and fitted with a good lock and key to prevent the weapons being meddled with.

A complete case should be equipped with cleaning gear, oil bottles, screw-driver, etc., and should have a partitioned section large enough to take a box of cartridges suitable to the weapon. The cleaning rod should be in a different compartment to the barrel, so as not to knock against it in transit.

Excellent leather pistol-cases fitted with a carrying handle and straps are now the usual equipment of a target shot, but some people prefer the French system of enclosing the polished oak or mahogany pistol-case in a separate leather cover.

Holsters are made in several patterns: the usual Army model, with a buttoned flap covering the whole of the weapon, the American "cowboy" holster, and the shoulder holster. Of all patterns of holsters the only one that is any use for wearing on service is the "Olive" or cowboy type. This is open at the top, has the side cut away to admit the finger direct to the trigger, and has a hole bored in the bottom to let out rain or dust. With this type of pistol holster, which is worn on a cartridge belt that hangs from left hip to right thigh, the hand falls perfectly naturally on the butt of the weapon—ready for instant drawing.

A revolver worn in this way is the best equipment for the mounted man, but it should be noted that if the barrel is over six inches long, it is uncomfortable and likely to bump on the saddle cantle when the holster is shifted to the riding position on the right buttock.

The shoulder holster is an excellent device for carrying a concealed weapon, and consists of a leather holster of the Olive type, suspended from a loop of leather, which fits over the left shoulder, and is further secured by a light strap passing round the back and over the breast under the waistcoat. In this manner a large revolver can be carried inconspicuously, but readily accessible. A certain amount of practice is needed before the art of drawing the weapon swiftly can be mastered.

Weapons not in use should always be kept in cases, not holsters.

A queer device of combining an electric torch and a revolver has been recently marketed; it is quite impracticable, but deserving of mention. A tube carrying a battery, a mercury switch, and a series of lenses, on one of which is marked a black spot, is mounted upon the barrel and parallel to its axis. At 12 yards: a circle of light about 2 feet in diameter, with a black spot about 4 inches wide in its centre, is thrown upon the target. This light only comes on when the pistol is held level and horizontally, and the tube is adjusted so that the bullet hits the centre of the black spot. The idea is excellent, but in practice it is found that the black spot wobbles all over the place, and at the moment of trigger pressure the circle of light leaps to one side in the most disconcerting manner. As a weapon for house defence it might prove useful at close range, but as an effective combination of an electric spot light aiming device it is of no real value.

So much time has been wasted by inventors on similar devices that it is worth mentioning, if only to save further expenditure of time and money on an unsatisfactory principle.

The Maxim silencer as used on rifles has been applied to pistols with varying degrees of success. The application of the device to pistols or revolvers is illegal in some countries, and very rarely applied in others. I have never seen weapons so fitted exposed for public sale, but have seen one or two privately adapted. In practice they are not very efficiently silent, and tend to render the weapon unwieldy and inaccurate.

The system of fitting special "target grips" or handles to pocket or police revolvers with small butts is an excellent method of improving the efficiency of the weapons as target arms, the larger grip giving a steadier and more effective purchase for the hand.

If it is desired to form a cheap but permanent target handle for a revolver, a good plan is to bind the whole butt, metal butt strap, wooden stocks, and all with ordinary electrician's insulating tape, till the whole is covered with three or four loosely applied layers. The weapon should now be thoroughly warmed in a tin box in front of a fire in order to soften the tape and render it properly pliant and adhesive. It should then be tightly gripped in the firing position, and held so until cooled down and "set." It is as well to cover the hand with loose talcum powder before gripping the hot weapon, as the rubber compound in the tape oozes out and adheres to the skin. Petrol or benzine will remove it.

This principle gives you a good non-slipping grip perfectly adapted to your own hand, as the tape moulds itself exactly to your grip. A handle so treated and afterwards varnished will stand years of wear. Weapons for use in the Arctic, where it is impossible to touch bare metal, should also be so treated.

Holster stocks or attachment stocks of any kind are not of much use as fitted to pistols and revolvers. They only turn a good pistol into a bad carbine. For some kinds of work a wire stock may be of use as fitted to a long-barrelled target pistol for hunting work, but on the whole it is better to carry a folding ·22 rifle than a hybrid weapon. The revolver is not meant for long-range work, and though good shooting may be done with the Service weapon up to 100 yards, and fatal effects obtained at 400 yards, its limit of efficient accuracy is about 75 yards, and its probably extreme range well under a mile. Some weapons, like the Mauser, are sighted for extreme ranges, but even used as carbines their accuracy at anything over 400 yards is exceedingly variable. It should never be forgotten that though a weapon may only be efficient up to 100 yards, it may kill at 400, and extreme care should be used that bullets never overshoot a butt or are fired "up into the air," regardless of where they may fall. Shooting at a moored barrel 250 yards out at sea, the firing-point being up on a cliff about 100 feet above sea-level, I found that bullets from an ordinary ·38 pocket hammerless Smith and Wesson went through both sides of the target—an efficient instance of the power of a small weapon a long range.

CHAPTER XIX

BISLEY SHOOTING

Pistol competitions are held at most rifle clubs, but the great national competition is the National Rifle Association week at Bisley. For those to whom Bisley is accessible for Saturdays, the N.R.A. have provided excellent accommodation for revolver shots.

The best way of enjoying Bisley is to be a member of the N.R.A., or an affiliated club, such as the "Middlesex," or the North London.

The standard range is the 20-yard stationary target, and the shot will find open-air shooting, with wind and light conditions, vastly different to shooting in a closed gallery with steady artificial light. Whatever the conditions of wind or light, always take the same amount of sight, and correct by aiming at a different spot on the target, rather than by taking a "coarser" or finer sight. Thus, if the shot strikes a little too high, or a little too far to the right, aim lower or more to the left of the bull to correct it for your next shot.

Before entering for a competition, fire a few practice cards to get used to the windage, light, and the "feel" of your pistol.

When cocking for single-action target work, touch the cylinder lightly with the forefinger, in order to make sure that it has come round into *complete* alignment with the barrel. The reason for this is that sometimes even with the best of weapons the cylinder does not come quite true, and the cylinder stop fails to engage in the slot provided for it.

This does not affect the safety of the weapon, but does interfere with its accuracy, as the bullet entering the barrel gets a shaving taken off its side.

This touch with the forefinger brings the cylinder round into locking position, even if the lifter does not, and after practice the trick becomes instinctive.

For Bisley a coarse front-sight and a good clear notch are advisable, particularly for use at the disappearing target.

This target is also at 20 yards range, and bobs up for three seconds, then disappears for three seconds.

You may not lift your pistol from the firing-rail before the target appears, and the weapon must be lowered again after each shot. Experts advise practice with a metronome in order to get into the way of the time if practising at a stationary target, but three seconds is ample time to get the shot off in.

Get used to the point where the target rises, and recognize the exact background in front of which it bobs up. Do not worry about its moving up and down, but imagine it to yourself as a stationary target *always there.*

If you do this you will find that when the target comes up you align on the bull naturally without any wavering about of the fore-sight to find the bull.

The 50 yards' range is the long range and the hardest. Here fine sights are an advantage, and slow deliberate shooting with heavier charges the rule.

The Olympic Games (when held) have some pistol-shooting events, and some on duelling or silhouette targets, but these do not correspond with Bisley competitions.

The handbook of the N.R.A., issued some weeks before the great Bisley week, contains details of all competitions, and the standard rules and regulations as given for 1914 are as follows:

B. REVOLVERS, PISTOLS, AND THEIR AMMUNITION.

169. No revolver or pistol may be used unless the makers have given a written guarantee to the Council that their revolvers or pistols, as the case may be, are suitable for use with the ammunition to be used with them, as set out in Regulations, paras. 172 and 175.

170. Every revolver or pistol and all ammunition is subject to examination by an officer appointed for the purpose before it may be used by a competitor, and is also subject to any test which the Council or Bisley Committee may prescribe.

171. No revolver or pistol may be used in any competition other than one of the class named in the conditions of shooting or in the special conditions applicable to any particular competition.

Revolvers and pistols allowed are classed as follows:

172. Any revolver complying with the following conditions:

Strength.—Must, in the opinion of the Bisley Committee, be suitable for military purposes, and capable of use with Government ammunition.

Weight.—Maximum, 2 lbs. 10 ozs.

Barrel.—Maximum length, 7½ inches.

Calibre.—Minimum diameter of bore, such as will carry ·44 ammunition.

Pull of Trigger.—Minimum, 4 lbs.

Sights.—Fore-sight: Must be firmly secured to the fore-sight lock or barrel, and incapable of lateral adjustment. The bead sight is allowed. The bead must not be less than ·08 inch in width and ·4 inch in length, nor must it be longer than any other portion of the sight. The narrowest part of the neck must not be less than half the diameter of the bead, and the sight must be generally of such solidity and strength as to be sufficient (in the opinion of the Bisley Committee) to withstand rough usage. *Back-sight:* To be without screw adjustment, and of some simple pattern of sufficient solidity and strength to withstand rough usage, in the opinion of the Bisley Committee. It may consist of a sliding bar or of leaves affixed to the barrel by hinges, or of interchangeable bar fitted into a slot; and it may have a plain open notch, or open square, or open half-circle or section of a circle, or of buckhorn pattern or a plain bar. The interchangeable bars fitting into a slot may be adjusted with the aid of a hammer, but such interchange of bars and such adjustment of bar, with or without the aid of a hammer, shall not be permitted at the firing-point.

Revolver Ammunition.—Only smokeless powder may be used. "Man-stopping" bullets may not be used. Competitors may use their own ammunition, which must conform to the following specification, viz.:

Weight of case: Maximum, 75 grains; minimum, 65 grains.

Weight of bullet: Maximum, 265 grains; minimum, 215 grains.

Velocity (average at 30 feet from muzzle): Maximum, 725 foot seconds; minimum, 600 foot seconds.

173. The Association will provide ·450 ammunition at the firing-point in exchange for ammunition tickets. These tickets can be purchased at the revolver ticket office, and each will entitle the competitor to six rounds.

174. No more than the actual number of rounds which the competitor is about to fire will be issued by the Association in exchange for ammunition tickets.

175. Any automatic or single-loading pistol, complying with the following conditions:

Calibre: Maximum, ·44 inches; minimum, ·22 inches.

Sights: Any, but must be affixed to the barrel.

Pull of Trigger: Single-loading, minimum, 4 lbs.; automatic, minimum, 4 lbs.

Pistol Ammunition.—Any ammunition suitable for use with the particular pistol used, except that the ammunition to be used with any particular pistol shall only be that which is covered by the guarantee given by the makers of the pistol, in accordance with para. 169.

Targets.

195. *Targets for Revolver Competition* are circular, mounted on square frames, and are coloured white, with a black bull's-eye. These targets are divided by rings into six divisions. The dimensions of the divisions and the marks scored for each hit in the several divisions are set out in Appendix No. 5. (Given below.)

Target Dimensions, etc.

Six Dimensions, etc.

Divisions.	Dimensions of Targets.		Scoring.
	Revolver, 50 Yards.	Revolver, 20 Yards.	
Bull's-eye	4 inches diam.	2 inches diam.	7 marks
1st Ring ...	6 " "	3 " "	6 "
2nd " ...	9 " "	4½ " "	5 "
3rd " ...	13 " "	6½ " "	4 "
4th " ...	18 " "	9 " "	3 "
5th " ...	24 " "	12 " "	2 "

K. REVOLVER AND PISTOL COMPETITIONS AT DISTANCES NOT EXCEEDING 100 YARDS.

432. All firing will be in the standing position.

433. Time-keepers are not allowed in individual competitions.

434. In vanishing competitions, the hand holding the revolver must be lowered to touch the table before each shot, and must not be raised until the target commences to rise.

435. Whilst loading, the muzzle must be held towards the target.

436. The revolver or pistol must not be loaded until the competitor is in position for firing, and the target has been run out to the butt.

437. Immediately after firing, and before quitting the table, every competitor must open his revolver and extract all

cartridge-cases; after this is done, and not until then, the target will be brought to the firing-point.

438. A competitor who, when opening his revolver and extracting the cartridge-cases after firing, holds his revolver in any manner other than in front of and clear of the table, and with the muzzle pointing to the ground in front of the target, may be excluded from further competition during the meeting. Any violation of the regulation contained in this paragraph is to be at once reported by the range officer, who is to warn the competitor concerned that he has been reported.

439. When firing at a disappearing target, the shots allowed are to be fired at six consecutive appearances of the target, and not more than one shot may be fired at each appearance.

440. When more than six shots are to be fired at one distance, they will be fired in series of six.

441. The second and each subsequent series of six shots is to be commenced as soon as the target is clear, and competitors must complete their firing at each distance, or retire from the competition, before quitting the firing-point.

442. At distances under 50 yards a new target will be given for each series of six shots.

443. Should a missfire occur, then: (*a*) If the competition is one in which there is no time limit, or in which there is a separate time limit for each shot or each appearance of the target, the competitor will continue to fire the remaining chambers, and will then place his revolver or pistol on the table. He will then at once reload as many chambers as may be necessary to complete his firing, which he will do forthwith, and without quitting the firing-point, receiving, if necessary, an allowance of time or of appearances of the target for the shot or shots to be fired proportionate to the time limit or to the number of appearances of the target allowed by the conditions of the competition. This time will be counted from the word "all clear" by the register keeper, which will be given as soon as the revolver is reloaded. Or (*b*) if the competition is one in which there is a time limit for a series of six or more shots, the competitor shall have the option of counting his score as it stands, or of cancelling the target and shooting over again at a fresh target. Under no circumstances, however, shall a competitor be allowed under this regulation to compete more than twice in respect of one entry.

444. Should there be from any cause an interruption in the shooting, the competitor must at once lay his revolver or pistol

on the table, muzzle towards the target, and will complete his shooting as soon as the target is clear.

445. Should the interruption be not caused by any fault on the part of the competitor, he is entitled to receive allowances as in the case of missfires.

446. At pool, when more than one competitor is present, they must fire in turn, and each must fire a complete series of six shots before quitting the firing-point.

447. When a pistol is used, the following additional regulations apply : (*a*) A pistol is to be held over the table pointing to the ground in the direction of the target, and may only be loaded whilst in that position. The pistol, still held in that position, may be cocked, or in the case of hammerless pistols, the slide taking the first cartridge into the barrel may then be operated ; after which the pistol, still held over the table, pointing to the ground in the direction of the target, must touch the edge of the table until the range officer or register keeper has given the words "ready—commence." (*b*) A wrist-rest attached to the heel-plate may be used, but it may not be attached to any part of the competitor or his clothing, nor may it touch the ground, or table, or any other support.

448. The trigger is to be tested whenever a score of forty or over has been made with a series of six shots.

Few pistol shots are in favour of the standard of pistol shooting and competitions as extant at Bisley, and there is no doubt that there is room for improvement ; but so long as the N.R.A. is the national authority on shooting, as distinct from the military authorities (whose ideas are, as as far as is known, equally hostile to either practice or progress with regard to pistol shooting), their rules are absolute.

Among noted revolver competitions are those of the Army Rifle Association, which limits itself to members of the Regular Army and Special Reserve on the Active List.

There is the Army Cup, a competition shot for by members abroad.

Practice 1 (*Right Hand*).—20 yards. A kneeling figure with three rectangles described therein: *inner rectangle*, 4 inches wide by 8 inches high; *intermediate rectangle*, 6 inches wide by 12 inches high; *outer rectangle*, 12 inches wide by 16 inches high.

The rectangles are to be invisible from the firing-point, and drawn so that their lower sides coincide with the bottom edge of the figure. Centre of lower side coinciding with centre of bottom edge of figure.

Six rounds in twenty seconds. Rounds not fired to be forfeited. Scores: inner rectangle 5 points, intermediate 4, outer 2, remainder of target 1. Highest possible score, 30.

Practice 2 (*Left Hand*).—As for Practice 1.

Practice 3 (*Right Hand*).—20 yards. Six new cardboard squares, coloured green, brown, or grey, 12 inches square. An interval of 3 feet between each target. Six rounds, one at each target, in fifteen seconds; unfired rounds forfeited. Score: 5 points for each hit; not more than one hit will be counted on any target. Highest possible, 30.

Practice 4 (*Left Hand*).—As for Practice 3.

Practice 5 (*Right Hand*).—20 yards. A revolver target, circular rings on a square card, subdivided as under:

20-yard card: Bull's-eye 2 inches in diameter; score, 7 points.

Ring	3	,,	,,	,,	,,	6	,,
,,	4½	,,	,,	,,	,,	5	,,
,,	6½	,,	,,	,,	,,	4	,,
,,	9	,,	,,	,,	,,	3	,,
,,	12	,,	,,	,,	,,	2	,,

Corners of card do not count. Six rounds in one minute. Rounds not fired to be forfeited. Highest possible score, 42.

Practice 6 (*Left Hand*).—20 yards. Same as Practice 5.

General Rules.—Dress, any. Position, standing—no rest or support, nor must the competitor's body be touching anything.

Revolver.—Any pistol, including automatic revolvers taking Government ammunition. Trigger-pull not less than 4 pounds.

Ammunition.—Smokeless; bullets, the Government conical or cartridges, S.A. Ball, Pistol Webley, Mark IV.

Place.—Any that may be considered convenient and safe for revolver practice.

Method of Conducting.—An officer, duly qualified in accordance with General Rule 26, must be present to superintend at the firing-point.

The officer superintending will carefully measure the 20 yards range, and will satisfy himself before the competitor commences a practice that the target is (Practices 3 and 4, targets are) free from shot marks and of the correct size, also that the rectangles in Practices 1 and 2 are accurately drawn. He will give the command "Fire" as soon as the competitor has loaded and is ready to commence the practice, and will time the practice from that command to the first sound of the whistle for "Cease Fire." The competitor will load and may aim before the command "Fire" is given. When the competitor has completed the practice, the officer superintending will proceed to the target and enter the number and value of the hits on the register. He will allow no signalling of shots.

The practices are to be fired in the order laid down above. During the match the competitor will remain at the firing-point.

Ties.—Ties in this match will be decided as follows; 1st, by the best score in Practice 2; 2nd, by the best score in Practice 1; 3rd, by the best score in Practice 4; 4th, by the best score in Practice 3; 5th, by the best score in Practice 6; 6th, by the best score in Practice 5.

The highest possible score for the Army Cup is 204. The scores were in 1913:

1. Lieutenant A. V. Pope, 14th Hussars—196.
2. Sergeant W. Plomer, 3rd K.R.R.—186.
3. Lieut.-Colonel W. Chitty, 119th Infantry I.A.—182.

The Revolver Thirty Match is the second A.R.A. event, and the object of the match is to select the best possible team to represent the Regular Army in the N.R.A. Revolver Team Match at Bisley.

First Stage—*Practices.*—Five preliminary practices will be fired by competitors at places where they are quartered. Conditions as follows:

Twelve shots at 20 yards, followed by twelve shots at 50 yards, either hand being used at the discretion of the firer. Stationary cardboard targets as described in Practices 5 and 6 of the Army Cup, which must be provided by competitors; these need not be sent to the secretary with the registers, but are not to be destroyed before August 1. One minute for each series of six shots at 20 yards, and one and a half minutes for each series of six shots at 50 yards.

The 50 yards target for the Revolver Thirty and other N.R.A. and A.R.A. events:

Bull's-eye	4 inches in diameter;	score,	7 points.
Ring	6 " "	"	6 "
"	9 " "	"	5 "
"	13 " "	"	4 "
"	18 " "	"	3 "
"	24 " "	"	2 "

Corners not to count.

Revolvers and ammunition must comply with N.R.A. regulations. Automatic revolvers may be used, provided they comply with the N.R.A. regulations as regards calibre, etc.

The standing position is to be adopted. No rest of any kind, natural or artificial, is allowed, nor may any competitor support himself against anything, or use more than one hand to steady his revolver. The whole of the arm with which the firing is carried out must be clear of the body.

Only one register may be filled in on one day by a competitor.

An officer superintending, who may belong to the same unit as the competitor, but must be otherwise qualified in accordance with General Rule 26, must be present at the firing-point during the whole of the firing. He will fill up and sign the register and will time the shooting at each distance. The competitor must declare before commencing to fire that he is about to shoot one of his five practices, which must then be completed without delay. The value of each shot must be entered on the register and the targets in ink. The register is to be posted by the officer superintending to the secretary on the day the practice is fired.

SECOND STAGE—*The Revolver Thirty.*—The Revolver Thirty will be selected by the committee on June 15, as follows: First fifteen in order of merit of the averages of the five preliminary practices in the First Stage, and the remaining fifteen at the discretion of the committee.

The selected Thirty will then meet together during the A.R.A. Annual Rifle Meeting, and compete for the prizes shown below. The conditions will be as follows:

Twenty-four shots at 20 yards, followed by twenty-four

shots at 50 yards, either hand being used at the discretion of the firer. Stationary cardboard targets as described in General Rule 27. One minute for each series of six shots at 20 yards, and one and a half minutes for each series of six shots at 50 yards. Revolvers and ammunition must comply with N.R.A. regulations. Competitors will provide their own ammunition. Any dress may be worn. Facilities for practice at Bisley by members of the Revolver Thirty on days prior to the match will be arranged for, competitors finding their own targets and ammunition.

Ties.—A tie for any one of the Jewels and the Money Prizes attaching to it, also for the N.R.A. Medal, will be decided, before leaving the range, by firing six shots at 50 yards, and, if still a tie, by repeating.

Special Rule.—Immediately on the conclusion of the match, the Captain of the Army team will select ten or other number of competitors, who will practise together as a team until the day of the N.R.A. Whitehead Cup match, under the supervision of a specially selected officer.

Prizes.—The three competitors making the highest aggregate scores will be awarded respectively the A.R.A. Revolver Gold, Silver, and Bronze Jewels.

The Bronze Medal of the National Rifle Association will be given to the competitor making the highest aggregate score who has not previously won this medal. Winners of this medal are entitled to compete for the Revolver Gold Medal at the National Rifle Association Meeting at Bisley.

The highest possible score is 336 points. In 1913 the highest scores were :

1. Lieutenant Egerton, R.E.—278.
2. Lieutenant Yeatman, S. Wales Borderers—277.
3. Sergeant-Major Kempster, 2nd Jersey Light Infantry—275.

CHAPTER XX

A SHORT COURSE OF ACTIVE SERVICE REVOLVER TRAINING

TRAINING for war should specialize upon treating its pistol-shooting as the use of an *arme de combat* rather than as a target weapon for ambidextrous use.

A revolver in competent hands is the deadliest of all modern small arms, and national efficiency in pistol-shooting is no less desirable than is our skill in musketry. An officer's personal connection with musketry ceases when he can train his men and control their fire. In combat the officer's weapon is the pistol, and he should be as skilled in its use as are the rank and file with their rifles.

It is hard to add another " course " to the modern educational system for war. Officers are already overburdened with general and specialized courses in everything, from " gas " to machine gunnery ; but when I consider the numbers of young officers whose ignorance of their one weapon has led them into the firing-line in an almost defenceless condition, I cannot but believe that it would be worth while in the interests of the nation were more time devoted to pistol practice. The importance of " double-action " practice has been slurred over by the target shots, and experts of this school frequently condemn its practice.

In time of war the military virtues of a revolver must be considered apart from its target capacities, and we find that the "double-action" principle has always had a following.

The "cowboy" school of novelists have much to answer for, for they have unconsciously grafted myths of the Western hemisphere on to revolver shooting in general, and it is from the West that we get the condemnation of the double-action.

Another tradition that has come from this school is the action of raising the pistol above the head before swinging down on to the target.

It has been assumed that this practice arose from the use in single-action revolvers of excessive loads designed for rifles such as the ·44-·40 Winchester. Actually the practice is considerably earlier, and arose with the early Colt muzzle-loading revolvers, in which *the arm was held perpendicularly while cocking, in order to prevent the exploded caps falling off the nipples and clogging the action of the cylinder.* Held properly in this way the exploded cap was disengaged and fell clear.

All officers whom I have consulted with regard to their practical experiences with revolvers in this present war are agreed upon the value of double-action work, and I can reinforce their modern testimony with the following quaint old letter addressed to Mr. R. Adams, the inventor of the Adams muzzle-loading revolver :

"Sir,

"In these days of warfare, any invention of improvements in firearms should be patronized and assisted, and with that view I write you this letter. I had one of your largest-sized Revolver Pistols at the bloody battle of Inkermann, and by some chance got surrounded by Russians. I then found the advantages of your pistol over Colonel Colt's, for had I to cock before each shot I should have lost my life, but with yours, having only to pull the trigger, I was able to shoot four Russians, and thereby save my life. I should not have had time to cock, as they were too close to me, being only a few yards from me, so close that I was bayoneted through the thigh immediately after shooting the fourth man. I hope this may be of service to you, as I certainly owe my life to your invention of the Revolver Pistol.

"I have the honour to be

"Your obedient servant,

"S. G. Crosse,

"88th Regt.

"Mr. R. Adams."

This Adams revolver was the weapon with which the East India Company armed some of its cavalry during the Mutiny. I have not been able to trace details of the institution of these weapons as a cavalry arm in this offshoot of the British service, but have come across references to an incident of the Mutiny reported in contemporary journals in which one hundred of our cavalry charged the rebels with a *revolver charge*, and in a few minutes left four hundred dead upon the field.

Perhaps some officer of the Indian Army will be able to furnish the information from his regimental traditions, in which case, if he will write to me, care of my publishers, I shall be deeply grateful.

It is difficult to lay down hard-and-fast rules for a course of training for military pistol-shooting, but elaboration of the following scheme will provide a course that can be easily combined with the existing training of young officers, and can be adapted according to the resources of the training area. The main points are that individual care and attention should be given by the officer-instructor to each young officer, and that practice should be frequent and not limited to the disposal of a meagre number of rounds of S.A.A. Pistol Webley.

An outline course of pistol-shooting should begin with :

1. *A Lecture.*—This should cover general particulars of the Service revolver and similar makes; the care of weapons; the limitations of their use; the fallacy of the superiority of small-bore automatics. The uselessness of the ordinary type of holster and the danger of lanyards fouling the hammer should be carefully pointed out, and officers encouraged to adopt the open Olive holster slung below the tunic in Western American fashion, for Service use. Every incident or anecdote that can be used to strengthen the interest of the class in revolvers and revolver-shooting as part of the *essential* knowledge of a competent officer should be brought in.

2. *Demonstration* of the correct grip of a revolver; method of loading and unloading; dismounting for cleaning, reassembling, etc. Never close a revolver with the hammer cocked. Correct position of holster. Smooth, light trigger-pulls. *Never point a pistol anywhere but up the range.* Explain the principle of instinctive sense of aim and touch, and the necessity for sticking to one type of weapon and not having continual changes. Lesson on aiming. Explain the absolute and unbreakable rule that whenever a pistol is taken in hand it must forthwith be inspected—*i.e., the action "broken" to see there are no cartridges in it.*

3. *Preliminary Practice.*—Inspect arms. Then proceed to aiming practice. This should be carried out with empty weapons, and the recruit taught to cock, level, and press his trigger the moment his sights align upon the bottom of the target. The instructor should stand behind an inspection disc with his eye to the aperture below the six o'clock bull. He should correct individual errors and mistakes, such as

snatching, canting, and flinch. Speed should be encouraged and insisted upon, but not so much as to disconcert or flurry the recruit. Within a short while he should be able to cock, aim, and fire within two seconds. Recruits should be instructed to continue this practice of snapping at an aiming mark in their spare time and indoors.

4. *Range Practice at Large Target.*—This should be carried out at close range—say 12 yards—and at large targets. Recruits should be accustomed to the shock of recoil, and encouraged to group their shots. In cases of flinch an excellent practice can be effected by loading the weapons with odd cartridges, some loaded, some empty cases. The trigger pull as a *squeeze of the whole hand* must be insisted upon. Speed of fire must also be encouraged, and any tendency to dwell upon the aim rebuked. The trigger should be pressed as the sights fall into alignment with the point of aim.

5. *Double Action Practice.*—Inspect arms. Clean arms. The class should have the value of double-action work for close action explained to them. Snapping practice with the double action. By continual use of the double action muscle is developed in the hand and arm, and much greater steadiness and rapidity of fire results when the arm is trained. Explain tendency of double action to throw high left when used right-handed, and high right when fired from the left hand.

6. *Double-Action Target Practice.*—A preliminary twelve rounds fired fairly slowly, and then strict insistence upon speed. This practice must be continued from day to day till a reliable passing-out standard of six shots in twelve seconds is attained. These shots to be within an 8-inch circle at 12 yards. Inspect and clean arms.

7. *Traversing and Bobbing Targets.*—These depend upon local facilities for their erection. A gravel-pit or chalk quarry is an ideal practice-ground. Counterbalanced targets may be made to bob up from a trench, or simple gravity targets running along a wire higher at one end than the other can be employed. An excellent variation of this can be used as a charging man: a frame target descending from a point in the quarry wall toward the firer along two parallel wires. The recruits should be taught first by empty snapping the amount of "swing" to allow for alternate bobbing right and left targets. Check tendency to use too much effort and haste when swinging on. Teach *low aim* at charging man. Aim always to be taken at centre line of body at traversing targets and figures.

8. *Duelling Practice.*—When the recruits have developed some facility at moving or bobbing figures, they can be set in pairs to fire at silhouette targets raised for three seconds for a single shot from a trench. In this three seconds the drawing of the pistol from the holster to the time of discharge is all included. The first to fire and hit wins. For rapid double action the figures may be raised for ten seconds, and the relative scores on time and hits *anywhere on the target* checked by the scale in the Appendix. This type of practice stimulates wholesome rivalry, and is excellent training for nervous shots.

9. *Duelling Practice* at traversing targets or alternating bobbers appearing at unknown points from the trench. This is run on lines as the above, and in every case the pistol should be returned to the holster and drawn again between single shots.

10. *Long-Range Practice.*—This applies to ranges over 20 yards and up to 50. The Service pistol (Webley) is usually issued sighted for this range, but target weapons are mostly sighted for the 20 yards distance. Speed should not be insisted upon, but too long dwelling upon the aim should be suppressed.

11. *Charging Practice.*—In this the recruits must run in Service kit over a chosen course, meeting concealed targets. They should practise the difficult art of reloading while running. A variation of this is to charge a trench and wire, firing at figure targets that rise from the trench.

12. *Left and Double Handed Shooting.*—It is obviously essential that an officer should be capable of using a revolver in either hand in case one arm is disabled. I am personally of the opinion that it is unnecessary to insist upon the recruits attaining an equal proficiency with both hands, as one continually finds that right-handed men have a right master eye, and the converse with naturally left-handed shots. Provided that a fair "instinctive" shoot can be made left-handed, using both single and double action, I would not insist upon an equal left and right test as part of the "passing-out standard." Double-handed work—that is, with a pistol in each hand—has a distinct military value, but only comes naturally to the ambidextrous. Shooting with both eyes open at two targets near to one another and in the same field of vision, it will be found that the point of aim or vision should be between the two; instinct will direct the fire of the right and left weapons to the real points of aim, the targets.

Such a course as I have outlined above can be carried out with other training, and can be so compressed that, provided the recruits are keen and fit, they can attain the requisite standard in ten to fourteen days.

It must be clearly understood that this does not turn them out as completely qualified expert pistol shots, but it gives them a training which will safeguard their own lives when in personal conflict with the enemy.

I do not believe in the efficacy of ·22 practice as part of an active service pistol course. All practice is useful, but I would prefer the recruit to use the full-weight Service pistol and Service ammunition as much as may be, and to practice snapping, drawing, and instinctive shooting with an empty pistol in his bedroom, rather than occupy his time with the ·22 weapon of different grip and no recoil.

The officer-instructor should be a keen man, and have a wide range of liberty in "how" he takes his recruits through and what type of targets he selects for them. I believe that where the terrain permits of it the use of full-size silhouette figures in neutral colours is by far the best training, and I would insist much on providing the element of surprise in the appearance or movement and direction of the target.

When space is restricted, good practice may be had at old bottles, jam tins—in fact, anything that may serve, and, with great caution, even shooting in the dark may be practised. The officer-instructor should be relieved of other duties in so far as is possible, and should devote much attention to the maintenance of ranges and the difficult task of squeezing more ammunition and various necessary materials out of the reluctant departments concerned.

The faults of the individual must be correctly diagnosed, elucidated, and corrected, and, above all, the indifferent shots should be encouraged. Lectures and cross-examination should be frequent, and questions provoked and answered. A complete record of all practices and courses should be kept, and besides the necessary "passing-out standard," a series of standards of higher excellence than the average should be maintained, so that the better the performance the more it redounds to the credit of the individual on his certificate of efficiency.

CHAPTER XXI

CLUBS AND RANGES

THE construction of revolver ranges is more the business of an architect or engineer than that of a pistol expert, but "wherever two or three persons are gathered together" and interested in shooting, a properly organized club and range is the first necessity. An outdoor range should, if possible, be 60 yards long from firing-point to target, but long-range work means the establishment of proper pit targets and shelters for markers—a moderately expensive undertaking. The best orientation for an outdoor range is to have the targets due north of the firing-point. The cheapest material for building safe butts, etc., is old railway sleepers and old rails, and it is far better to adopt and adapt such material as can be found locally than to invest in expensive iron frames and carriers that have to be specially ordered. There is a distinct supposition on the part of many writers of pistol books that shooting men are necessarily wealthy, and it is common to find advice about expensive refinements, electrical fittings, and the like. My own experience tends to rather the opposite point of view, and I do not find shooting confined to, or greatly supported by, the ultra-wealthy, and except as offshoots of various rifle and miniature rifle clubs, there are no revolver clubs of note.

A good range in an old quarry, chalk, or gravel-pit, with a covered shed firing-point, is about as good an arrangement as is necessary. Up to 20 yards no marking arrangements are needful, as the cards can be run down to the firing-point on carriers supported by parallel wires, and actuated by plain rope and pulley winding gear. Moving, traversing, and up-and-down targets can also be actuated from the firing-point. The result of each shot can be seen by anyone of average eyesight, and the whole expensive paraphernalia of pits and markers' shelters, etc., is thereby avoided.

Patent target devices that run on complicated systems of troughs and pulleys are seldom satisfactory. A plain pair of taut wires to carry a sheet-steel target carrier, and pulleys made of old perambulator wheels (the tyre groove takes a rope nicely), set in plain frames of ordinary gas piping, or the wooden firing-point counter, make the simplest and most efficient range equipment.

Indoor ranges of 20 yards or less need the same equipment, and the back, roof, and sides of the range should be plated with sheet-iron thick enough to stop or shatter the ·38 automatic cartridge, which is the most penetrating pistol charge used. If the range is underground, only the butts and roof need protecting from accidental discharge.

The whole face of the butt should be steel or iron-plated, and behind each target should hang a heavy plate of ½-inch steel about 24 inches by 24 inches square. This receives the maximum battering, and can be easily replaced when worn out.

The lighting of the targets should be from gas or electric lights, about 2 feet in front of them, and about 18 inches below their base level. As splashes of lead and ricochets easily destroy the lamps or mantles, these should be partly enclosed in tin reflectors, so focussed as to throw the maximum of light upon the targets, but without exposing more than possible of the lamps to splashes. A good plan is to erect a thin board or plate-shield just under the targets, and between the lamps and the steel plate. This shield stops most of the splash.

The lighting arrangements are shielded from the firing-point by timber faced with heavy plate, and at least two lamps are used to light each target. A point to remember with the modern metal filament lamp is the need of minimizing all possible vibration, and the lamp settings should be unattached to the protecting plate, and, if possible, slung upon a loose board that does not suffer from the shock should a bullet hit the protecting steel sheeting.

The distance from firing-point to target must be exact, and the floor firm. Concrete covered with cocoanut-matting affords a good, firm standing ground.

At Gastinne Renette's and most Continental galleries of note each shooter fires through a special loop-hole or casement window of his own. This system partitions off the shooters and renders the range safer and less noisy. At Gastinne's the silhouette targets are automatic, electrically indicating, but the bull's-eye cards are called by the marker in the usual way.

Most of the patent electrically registering targets will not stand the smashing impact of revolver bullets, and soon go out of action. I have not yet found any of these devices that will work properly with anything more powerful than the ·22 ammunition, and they are mostly expensive and unreliable.

The most modern of all ranges is the special cinematograph or living picture range. This needs an expensive installation and several attendants, the mechanism complete costing some five hundred pounds. For this the usual cinematograph picture, or specially made films, are projected on to a paper screen. There are three of these screens, one behind the other acting on the roller-blind principle, one moving laterally, the other two vertically.

When a shot is fired the sound of it automatically stops the progress of the film, and the picture stops rigid for a second, showing the shot as a bullet-hole or spot of light, for behind the screens there are powerful illuminants.

The sound of the shot also sets in action the rollers of the paper screens, and after a second's delay all three screens move half an inch in different directions. This automatically closes the puncture, and prevents the light behind being seen. The film again starts up, and the picture progresses till arrested by a second shot.

Practice at these targets is amusing and instructive, particularly for the perfecting of rapid-fire work, but as the first report stops the screen, the device has its limitations, and is only useful for noting the effect of the first snapshot. The other shots only take effect upon a stationary target.

In conclusion it may be said that, after all, range practice is not essential, and that a good deal may be learnt by outdoor work by sea or stream, by wood or field. Anything will do for a target, and, provided the background is safe, a great deal of useful and enjoyable practice may be obtained.

Pistol shooting is fine sport and good fun. It needs nerve and skill and is not an expensive hobby, and if this little work does anything to assist those who love pistols, it fulfils the author's humble purpose.

CONCLUSION

AUTHOR'S NOTE

In writing a book of this nature, the author is bound to glean much from the work of others who have preceded him. I cannot include a complete bibliography, but I desire to express my indebtedness to the works of Mr. W. Greener, Mr. Walter Winans, Mr. C. W. Sawyer, Mr. Himmelwright, Mr. McGuire, the editor of that excellent American magazine, *Outdoor Life*; and many others. I also beg to thank Mr. Evans, of Wolverhampton, and Mr. Allen, of the Façade, Charing Cross, for the use of some photographs of antique arms. During my researches it has been a constant pleasure to me to find how well my inquiries have been received by leading gunsmiths and firms dealing in antique weapons; they have readily placed at my disposal any information which they possessed, and have manifested a keen desire to help me in the compilation of this work.

I desire to thank them and all those who have given me encouragement to proceed with it. The times have not been propitious, and it has been peculiarly hard to continue a work of this nature during the exigencies of military service in the field. This must serve as apology for the manifest shortcomings of the work, and I can only hope that in later editions I may be able to expand and improve a book that I hope may some day be the standard authority on pistols and revolvers.

HUGH B. C. POLLARD,
Captain.

War Office,
London, 1916.

APPENDIX I

Table I

COMPARISON OF BORE AND CALIBRE

Bore was originally ascertained by the number of balls to a pound of lead.

Bore or Gauge.		*Diameter of Bore (In.).*	*Bore or Gauge.*		*Diameter of Bore (In.).*
12	..	·729	32	..	·526†
13	..	·710	33	..	·520
14	..	·693	34	..	·515
15	..	·677	35	..	·510
16	..	·662*	36	..	·506
17	..	·649	37	..	·501
18	..	·637	38	..	·497
19	..	·626	39	..	·492
20	..	·615	40	..	·488
21	..	·605	41	..	·484
22	..	·596	42	..	·480
23	..	·587	43	..	·476
24	..	·579	44	..	·473
25	..	·571	45	..	·469
26	..	·563	46	..	·466
27	..	·556	47	..	·463
28	..	·550	48	..	·459
29	..	·543	49	..	·456
30	..	·537	50	..	·453
31	..	·531			

Bullets to fit these bores should be ·02 less in diameter.

* 1 oz. spherical ball. † ½ oz. ball.

TABLE II

Small Bore Gauges.			*Diameter of Bore.*	*Size of Bullet.*
51·05	..	..	·450	·430
54·61	..	..	·440	·420
58·50	..	..	·430	·410
62·78	..	..	·420	·400
67·49	..	..	·410	·390
72·68	..	..	·400	·380
78·41	..	..	·390	·370
84·77	..	..	·380	·360
91·83*	..	..	·370	·350
99·70	..	..	·360	·340
108·49	..	..	·350	·330
118·35	..	..	·340	·320
129·43	..	..	·330	·310
141·95	..	..	·320	·300
156·14	..	..	·310	·290
172·28	..	..	·300	·280

* Kentucky "pea rifle" size.

APPENDIX II

EUROPEAN REVOLVER CARTRIDGES AND CHARGES

Calibre.	*Powder (Grs.).*	*Bullet (Grs.).*
·297/230 (short) ..	3	38
·297/230 (long) ..	5	38
·295/250	7	56
·295	10	80
·300	10	80
·320 (short)	6	80
·320 (long)	9	80
·360 No. 5	14	134
·380 (short)	10	124
·380 (long)	15	146
·450	13	225
·455	18	265

APPENDIX III

REVOLVER CARTRIDGE LOADS (AMERICAN)

Cartridge.	*Black Powder (Grs.).*	*Bullet (Grs.).*
·22 short rim fire	4	29
·22 long rim fire	5	29
·22 long rifle rim fire ..	5	40
·25 rim fire	10	67
·32 short	6	80
·32 long	13	90
·41 Derringer	10	180
Centre Fire.		
·32 short Colt	9	80
·32 long Colt	12	90
·32 S. and W. short ..	9	85
·32 S. and W. long ..	13	98
·32 Colt police positive ..	13	98
·32-·20	20	100
·38 short Colt	18	130
·38 long Colt	19	150
·38 S. and W. short ..	14	145
·38 S. and W. special ..	21	158
·38 Colt special	21	158
·38 Colt police positive ..	14	150
·38-·40	40	180
·41 short Colt	14	130
·41 long Colt	21	200
·44 S. and W.	26	246
·44 Russian	23	246
·44-·40	40	217

Note.—With small rim-fire cartridges the heavy charge of fulminate also acts as a most powerful propellant.

APPENDIX IV

AUTOMATIC CARTRIDGES

Cartridge.	*Weight of Bullet in Grains.*	*Length of Barrel.*	*Initial Velocity (Feet per Second).*	*Penetration of 1 in. Pine Boards 1 in. apart at 20 Yards (Inches).*
·25 or 6·35	48	2	650	4
·32 or 7/365 mm.	75	3½	900	7
·35	—	—	—	—
·380	—	3¾	1030	8
·38	130	6	1100	—
·45 Colt	200	5	816	5¼
·45 Colt U.S. Govt.	230	5	830	—
·455 Webley ..	220	—	710	—

APPENDIX V

TABLE OF MODERN STANDARD REVOLVERS AND AUTOMATIC PISTOLS, WITH THEIR CALIBRES, BARREL LENGTHS, NUMBER OF SHOTS, AND WEIGHT

Model.	*Calibres.*	*No. of Shots.*	*Length over All (In.).*	*Length of Barrel (In.).*	*Weight (Oz.).*
COLT GROUP.					
Single action army (frontier and Bisley models)	·32	6	10¼	4¾	37
	·38		—	—	—
	·44		11	5½	—
	·45		13	7½	—
Pocket positive ..	·32	6	6½	2½	—
			7½	3½	16
			10	6	—
Police positive ..	·32	6	8¼	4	—
	·38		9¼	5	20
	—		10¼	6	—
Police positive target	·22	—	10½	6	22
	·32	—	10½	6	19
Army special ..	·32	6	9¼	4	33
	·38	—	10¼	5	33
	·41	—	11¼	6	35
New service ..	·38	6	9¾	4½	40
	·44		10¾	5½	42
	·450		12¾	7½	44
	·455		—	—	—
New service target	·455	6	12¾	7½	42
Colt automatic pistol (military)	·45	—	—	—	—
	·455	7	8½	5	39

Model.	*Calibres.*	*No. of Shots.*	*Length over All (In.).*	*Length of Barrel (In.).*	*Weight (Oz.).*
Colt automatic pistol (military) ..	·38	8	9	6	38
Colt automatic pistol (pocket models) ..	·38	7	7½	4½	32
	·380	7	6¼	3¾	23
	·32	8	6¾	3¾	23
	·25	6	4½	2	13
WEBLEY GROUP.					
Webley pocket (W.P.)	·320	6	7	3	17
Webley pocket (W.P.) (hammerless) ..	·320	6	7	3	18
Mark III. (pocket)	·38	6	7	3	19
			8	4	20
			10	6	21
Webley Mark V. (service revolver) and previous models ..	·455	6	9¼	4	35
			11¼	6	36
Webley W. and G. (Officer's model) ..	·455	6	11¾	6	40
Webley W. S. Bisley target revolver ..	·455	6	9¼	4	36
			11¼	6	38
			12¾	7½	39
Webley W. G. target revolver	—	6	11¾	6	40
	—	—	13¼	7½	42
Webley automatic (Navy model) ..	·455*	7	8½	—	36
Webley automatic (foreign military model)	9 mm.*	8	8	—	32
Webley automatic high-velocity ..	·38*	8	8	—	31
Pocket model ..	·380	7	6¼	—	18
Pocket model (M.P.)	·32*	8	6¼	—	18
Pocket model (hammerless)	·25*	6	4¼	—	9½
Pocket model (hammer)	·25*	6	4¾	—	11

* Weights less magazine.

Model.	*Calibres.*	*No. of Shots.*	*Length over All (In.).*	*Length of Barrel (In.).*	*Weight (Oz.).*
SMITH AND WESSON GROUP.					
Pocket models with breakdown action	·32	5	3 3½	— —	12½ —
Ditto	·38	5	3¼ 4 6	— — —	17 17¼ 19
Pocket models with solid frame swing out action ..	·22	7	3 6		10¼ 11½
Ditto	·32	6	3¼ 4¼ 6	— — —	17¼ 18½ 20
Military models, 1902, 1905, 1908	·38	6	4 6	— —	— —
Military models, solid frames	·44	6	6½	—	—
Pocket hammerless breakdown action	·32	5	3 3½	— —	14½ 15
Ditto	·38	5	3¼ 4 6	— — —	17 17¼ 19
Automatic	·35	7	—	—	—
Target pistols ..	·22	1	10	—	—
SAVAGE GROUP.					
Pocket automatic ..	·32	10	6½	3¾	19
,, ,,	·380	9	7	4¼	21
BROWNING GROUP.					
Pocket model ..	·25	6	4½	2	14
,, (old type)	·32	7	6½	4	24
,, (new type)	·32	7	6	3½	20
,, (new type)	9 mm.	6	6	3½	20
Military model ..	9 mm. (long)	7	8	5	33

APPENDIX VI

TABLE SHOWING COMPARISON OF BALLISTICS OF SMOKELESS REVOLVER CARTRIDGES AND THE SAME CALIBRE AUTOMATIC PISTOL LOADS

	Bullet (Grs.).	*Muzzle Velocity in Feet per Second.*
·455 Service	265	660
·455 target "naval bullet" ..	225	710
·450	225	640
·380	124	600
·320	80	550
·455 Webley naval automatic ..	224	710
·450 Colt automatic	200	830
·38	130	1,000
9 mm. Browning	110	950
·380	95	950
·32	75	900
·25	50	650

These figures vary slightly according to the make of ammunition used, as the makers have slightly different loads of varying kinds of smokeless powder.

Muzzle velocities are apt to be misleading if taken as a criterion of efficiency, and the figures of loss of energy in the large-calibre automatic pistol loads when the bullet has travelled fifty yards show an excessive drop:

·455 Service revolver at 50 yards has a velocity of 714 feet per second and an energy of 300 foot pounds.

·455 Automatic pistol at 50 yards has a velocity of 680 feet per second and an energy of 230 foot pounds.

·450 Colt automatic pistol at 50 yards has a velocity of 780 feet per second and an energy of 270 foot pounds.

The smaller calibres correct this loss thus:

·380 revolver at 50 yards has a velocity of 628 feet per second and an energy of 109 foot pounds.
·380 Automatic at 50 yards has a velocity of 850 feet per second and an energy of 152 foot pounds.
·38 Automatic at 50 yards has a velocity of 1,035 feet per second and an energy of 304 foot pounds.

This would seem to give the highest value to the long ·38 automatic cartridge, but in point of fact the smaller calibre and lesser diameter of the bullet lose so much "shocking power" that the ·45 calibres are much more effective.

APPENDIX VII

NAMES OF THE PRINCIPAL PARTS OF A FLINT-LOCK PISTOL

BOLTS—Pieces of steel that push through the barrel loops to fasten the barrel into the stock.

BRIDLE—The curved piece of steel which holds the scear and tumbler in position against the lock plate.

CAP—Tip of stock or covering for worm of ramrod.

CHAIN OR SWIVEL—The little catchpiece joining the neck of the tumbler and the mainspring. This is probably so called from the old wheel-lock, when this member often was an actual link or two of chain.

CHAMBER—The centre or principal tube in the breech screw. The *antechamber* is the smaller tube leading from touch-hole to chamber.

COCKSCREW—The screw that tightens the jaws to hold the flint.

ESCUTCHEONS—Pieces of metal or silver to prevent the bolts wearing the stock; also applied to the shield on which crest or monogram was engraved.

FACING OF FRIZZEN—Part of pan cover, or frizzen, which strikes fire from the flint.

FALSE BREECHING—The part where the breech of the barrel hooks in before the barrel can be fastened to the stock.

FALSE BREECH SCREW—The screw which secures this to the trigger plate.

FEATHER SPRING—The frizzen, or pan-cover spring.

FENCE—The curved guard of the pan that arrests the fall of the hammer.

FRIZZEN BRIDLE—The part that the frizzen extension works in.

GUARD—The bow which defends the triggers.

JAWS—The lips of the cock or hammer that hold the flint.

LOOPS—The eyes that hold the pins or bolts that secure barrel to stock.

MAINSPRING—The spring that works the tumbler and cock on hammer.

NIPPLE—The projection of the touch-hole on which a cap is placed.

PIPES—The bands on the stock to receive the ramrod.

SCEAR—The part which engages with the notches on the tumbler and holds the hammer at half full cock.

SCEAR SPRING—The spring that presses the scear into position.

SCROLL TRIGGER OR SPUR TRIGGER—An extra extension of the trigger-guard to steady the hand.

SIDE NAIL—The screw that holds the lock in position.

TAIL—The arch, neck, or extension of the frizzen.

TRIGGER SPRINGS—Small springs invented by Joseph Manton that keep the trigger always pressed against the scear.

TUMBLER—The moveable centrepiece of the lock that moves with and directs the hammer or cock.

TUMBLER SCREW—The little screw that fastens the hammer to the tumbler pin.

WORM—Screw end on a ramrod for drawing wadding.

Note.—On the flint-lock weapons the hammer, as we understand it to-day—that is to say, the falling member that held the flint—was termed the " cock "; the frizzen or steel or flash-pan cover was termed the " hammer." Later, when the percussion system came in, the cock was called the " striker," and gradually has come to be known as the " hammer."

Throughout this book I have called it the " hammer," as this name is less misleading to modern minds. When reading old books, it must be remembered that the term applies to the " flash-pan cover."

Frizzens and springs were easily worn out and often replaced, and nearly every gunmaker had his own idea about how a frizzen should be made and fitted so as to give good fire and waterproof protection to the priming. The test of a good frizzen was that it should " shut down dull and fly back smart." Frizzens could be " new steeled," that is to say, the steel plate welded on to the front of the frizzen renewed, but in most cases it was simpler to put in a new " hammer," and gunmakers always encouraged this process.

Colonel Hawker (1825) says:

"There are two ways of putting a 'hammer' (frizzen) in motion—one with a wheel in a feather spring, the other with a bridge there over which a wheel in the hammer is made to run. Durs Egg and Mr. Manton usually make the first; Mr. Joseph Manton the other. This is a good way of determining the maker of 'Manton' locks — only the Joseph Mantons being of great value to the collector; the other Manton—John Manton—was his rival and brother, and was once foreman to the celebrated Twigg."

The pan in flint-locks is placed low down below the touch-hole in order that the first flash of the priming fires the charge. In pieces where the touch-hole is at the bottom of the pan, the priming had all to burn away first, before the discharge occurred.

Flints should be put in flat side upwards, and held in a piece of leather. If they do not force the frizzen back while in this position, they may be reversed, which makes it strike lower. In the old days, flints were of special sizes, and had to be very carefully adjusted by packing more or less leather round them to give the best strike.

APPENDIX VIII

LIST OF LONDON GUNMAKERS IN 1825

This is from Lieut.-Colonel P. Hawker's book "Instructions to Young Sportsmen," fourth edition, which deals largely with the question of flint-lock as against the new "detonator" or percussion system.

Anderson, Robert.
Baker, Ezekiel.
Bates, John.
Barnett and Son.
Barton, J.
Beckwith, W. A.
Beddowes, John.
Biven, J. A.
Blake, J. A.
Blanch, John.
Bond, E. J.
Bond, William.
Bowstead, F.
Brander and Potts.
Egg, Durs.
Egg, Joseph.
Enty, John.
Forsyth and Co.
Goff, Daniel.
Green, William.
Grierson, Charles.
Gulley, Joseph.
Hall, Collinson.
Harding, James.
Hill, John.
Homer, Thomas.
Jackson, Richard.
Jackson, George.
Lacy and Witton.
Lombard and Butter.
Leigh, J.
London, William.
Manton, John, and Son.
Manton, Joseph, and Sons.
Moore, Charles.
Mortimer, Thomas.
Mortimer, Thos. J.
Nock, S.
Nock, Samuel.
Parker, William
Parkin, Thomas.
Pritchett, R. E.
Purdey, James.
Rea, T. A.
Reynolds, Thos.
Ridley, Thos.
Ridley, William.
Rivière, Isaac.
Sherwood, J.
Smith, Samuel.
Standenmeyer, S. H.
Stevens, Thomas.
Sturman, Philip.
Tatham, Henry.
Turney, J.
Webster and Co.
Wilbraham, George.
Wilkinson, James, and Son.
Wilson, Alex.
Wilson, William.
Wright, Robert.

Barrel-makers.

Lancaster, Chas.
Fullerd.

(Of the above names, Purdey, Blanch, and Lancaster are still in existence as first-class gunmakers to this day.)

APPENDIX IX

LIST OF SOME BRITISH GUNMAKERS

This list is compiled from various sources, such as Directories, Patent Records, books, etc., and gives merely the date at which a certain maker was at work. It does not necessarily mean that this was the first decade of his existence, and is compiled merely as a guide to the approximate dating of old weapons.

Joseph Adams, Birmingham	1770
Thos. Archer, B'm'gham	1780–1812
Thos. Addis, London	1632
J. Aspinal, Birmingham	1812
S. Anens	1690
Thos. Allport, Birmingham	1812
Jacob Austen, B'm'gham	1689
Robt. Bagnall, B'm'gham	1770
Barber, London	1760
Bunney, London	1720
Barbar, London	1705
Barbar and Harris, Birmingham	1770
Thos. Bays, London	1785
Isaac Bissel	1780
Wm. Bourne, B'm'gham	1689
John Borle	1770
Bidet, London	1790
Chas. Byrne	1772
Philip Bond, London	1776
Blair, Birmingham	1790
Wm. Bond	1812
Brander, London	1637–1850
Thos. Bridgens, Birmingham	1770
J. and R. Brooks, London	1686
Thos. Brielton	1770
R. Burrows, London	1632
Joseph Bunney, Birmingham	1770–1812
John Bagley, B'm'gham	1812
Thos. Bate, B'm'gham	1812
John Bingham, Birmingham	1812
Blair and Co., B'm'gham	1812
R. Bolton, Birmingham	1812
Richard Boot, Birmingham	1812
Birchett	1700
Clemson, Salop	1740
Cardiffe	1682
Collumbell, London	1743
James Cash, B'm'gham	1801
Collier, London	1818
Thos. Crodler, London	1661
Thos. Capner, Birmingham	1812
Wm. Cheshire, Birmingham	1812

Cooper and Banks, Birmingham 1812
Wm. Dawsten, London 1632
John Dunn, B'm'gham 1770
Dylaney, London .. 1700
Drury, Liverpool .. 1780
Daniels, Cross and Co., Birmingham 1812
Davidson Duncan .. 1812
Wm. and Sam Dawes .. 1812
John Dutton 1812
Mabson, Labron, Dunderdale 1812
Edge, London 1759
John Eginton 1801
B. Egg 1750
Durs Egg, London 1785 & later
Field, Aylesbury ..
Freeman, London .. 1700
Wm. Guice, B'm'gham 1770
Wm. Graves, London .. 1632
Thos. Galton, B'm'gham 1780
John Green 1775
Griffin, London .. 1796
Sam Galton, B'm'gham 1812
Tom Gill, Birmingham 1812
Ben Griffiths, B'm'gham 1812
A. Hall, London .. 1664
Gilbert Hadley, Bristol 1741
H. Hadley, London .. 1789
Thos. Hadley, B'm'gham 1770
Abraham Hill, London 1664
Jos. Hicking, B'm'gham 1770
John Hunt 1770
Lord and Hutchinson ..
Heeley, Birmingham .. 1790
Hutchinson, Dublin .. 1775
Hampton, Birmingham 1812
Heathcote and Evans, Birmingham 1812
Thos. Henshall, Birmingham 1812
Wm. Howell, B'm'gham 1812
Geo. Jackson .. ..
John Jonas .. ..
Wm. Jones .. ..
Wm. Jover 1775
Koster 1620
T. Ketland, B'm'gham 1750-1829
Ketland and Walker, Birmingham .. 1750-1829
Ketland, etc., Birmingham 1750-1829
Thos. Lane, B'm'gham 1770
D. McKenzie, London 1720
John Manton, London 1797
Joseph Manton, London 1766-1835
Mewis and Moseley, Birmingham 1790
Mortimer, London .. 1789
Moore, Birmingham .. 1689
H. Martin Muler, London 1685-89
Sam Mayo, B'm'gham 1812
Henry Morris, B'm'gham 1812
Joan Moseley, B'm'gham 1812
Thos. Moxham, Birmingham 1812
Henry Nock, London .. 1775
Samuel Nock, London 1812
Nock, Jover and Green, London .. 1775-80
Newton, Grantham .. 1776
John Norcott, London 1632
John Onions, B'm'gham 1770
John Osborne, B'm'gham 1770
Joseph Oughtom, Birmingham 1780
Page, Norwich 1776
Parkes, Birmingham .. 1770
Parr, London 1750
Puckle, London .. 1718
Parrett, Salisbury .. 1760
John Pasmore, London 1640
Wm. Perry, B'm'gham.. 1780
Prosser, London .. 1770

Thos. and Chas. Probin, Birmingham .. 1770-1812
Radoc, Norwich .. 1590
Richards, London .. 1700
Thos. Richards, Birmingham 1770
John Richards, Birmingham 1770-1812
Westley Richards, Birmingham 1820
Redfern, Birmingham .. 1790
Henry Rowland, Birmingham 1632
R. Rowland, B'm'gham 1718
Ryan and Watson, Birmingham 1780
Reeves, Birmingham .. 1812
Reddell and Co., Birmingham 1812
Thos. Reynolds, Birmingham 1812
John Richards, Birmingham 1812
Theo. Richards, Birmingham 1812
W. J. Rolfe, B'm'gham 1812
James Rose, B'm'gham 1812
Robinson, London ..
Wm. Staples, B'm'gham 1770
Sharpe, Birmingham .. 1730
Joseph Simmonds, Birmingham 1812
Shirts, London .. 1795
Smart, London .. 1800
John Southall, Birmingham 1770
John Sharpe, B'm'gham 1812
John Smith, B'm'gham 1812
Joseph Spencer, Birmingham 1812
Joseph Such, B'm'gham 1812
L. Tanner, London .. 1795
Templeman, Salisbury, Wilts 1789
Tatnam, London .. 1800
Ed. Taylor, B'm'gham 1770
E. Tilly, London .. 1690
Thomas, Birmingham .. 1790
Tow, London 1789
Twigg, London .. 1782
Truelock, London .. 1667
S. Turner, London .. 1776
W. Thompson, Birmingham 1812
Ed. Timmings, Birmingham 1812
John Turner, Birmingham 1812
Wm. Walsingham, Birmingham 1770
John Watson, London 1632
John Waters, B'm'gham 1770
Tom West, B'm'gham 1689
John Welch, B'm'gham 1770
Richard Weston, Birmingham 1689
Willets, Birmingham .. 1769
Wogdon, London .. 1770
J. Whateley, Birmingham 1770
Robert Wheeler, Birmingham 1770
Richard Wilson, Birmingham 1770
Ben Williams, B'm'gham 1770
Walker, Birmingham .. 1790
Wicksted, Fenchurch St., London .. ..
Yeomans 18
York 18

APPENDIX X

SOME EUROPEAN PISTOL MAKERS

Italian.

Lazaro Lazaroni, Venice	1640
Lazarino Comminazzi „ Commazzo „ Cominazo } *	1600-1780
Laro Zarino	
Lazaro Lazavino	
Gio Maria Aquisti	1630–1720
Matteo Badile	*circa* 1680
Gio Bat Francino	*circa* 1680
Antonio Bonisolo	*circa* 1720
Bartolomeo Cotel	*circa* 1740
Tohandz, Brescia	*circa* 1750
Postindol, Spezzia	*circa* 1750
Geronimo Motto	*circa* 1750
Giocatane	*circa* 1750

Spanish.

Lasinto Laumandrew	1739†
Morino	1745
Martin Marchal	1780
Sahagom	1680

French Gunsmiths.

Claude Thomas, Epinal	1623
Simorim, Luneville	1627
Pierre Baroz, Paris..	1780
Bouillet, Paris	1715-1774

* A family of Italian pistol makers.

† Revolver in Tower of London Collection.

FRENCH GUNSMITHS—*Continued.*

Boutet, Versailles	1780–1825
Brezollaine, Paris	1800
Brezol, Caen	1820
H. Rénier, Paris	1780
I. Rénier, Paris	1750
Lamarre, Paris	
Duboirs, Paris	
Le Faucheux, Paris	1820
Gastinne Renette, Paris	1850
Boutet, Marseilles .. .. .. *circa*	1760
Mercier, Liège	
Pavly, Genève	

APPENDIX XI

SCORING SCALE FOR REVOLVER SHOOTING AT FIGURE OR BLANK TARGETS OR WAX-BULLET PRACTICE AT INDIVIDUALS

PROPORTIONAL VALUES FOR SPEED AND ACCURACY.

Speed.		*Shots on Target.*					
Secs.	*Fifths of a Second.*	**6**	**5**	**4**	**3**	**2**	**1**
20		120	100	80	60	40	20
19	4/5	121	101	81	61	40	20
19	3/5	122	102	82	61	41	20
19	2/5	124	103	82	62	41	21
19	1/5	125	104	83	62	42	21
19		126	105	84	63	42	21
18	4/5	128	106	85	64	43	21
18	3/5	129	108	86	65	43	22
18	2/5	130	109	87	65	43	22
18	1/5	132	110	88	66	43	22
18		133	111	89	67	44	22
17	4/5	153	112	90	67	45	22
17	3/5	136	114	91	68	45	23
17	2/5	138	115	92	69	46	23
17	1/5	140	116	93	70	47	23
17		141	118	94	71	47	24

Speed.		*Shots on Target.*					
Secs.	*Fifths of a Second.*	6	5	4	3	2	1
16	4/5	143	119	95	71	48	24
16	3/5	145	120	96	72	48	24
16	2/5	146	122	98	73	49	24
16	1/5	148	123	99	74	49	25
16		150	125	100	75	50	25
15	4/5	152	127	101	76	51	25
15	3/5	154	128	103	77	51	26
15	2/5	156	130	104	78	52	26
15	1/5	158	132	105	79	53	26
15		160	133	106	80	53	27
14	4/5	162	135	108	81	54	27
14	3/5	164	137	110	82	55	27
14	2/5	167	139	111	83	56	28
14	1/5	169	141	113	85	56	28
14		171	143	114	86	57	29
13	4/5	174	145	116	87	58	29
13	3/5	176	147	118	88	59	29
13	2/5	179	149	119	90	60	30
13	1/5	182	152	121	91	61	30
13		185	154	123	92	62	31
12	4/5	187	156	125	94	63	31
12	3/5	190	159	127	95	63	32
12	2/5	194	161	129	97	65	32
12	1/5	197	164	131	98	66	33
12		200	167	133	100	67	33
11	4/5	203	169	136	102	68	34
11	3/5	207	172	138	103	69	34
11	2/5	211	175	140	105	70	35
11	1/5	214	179	143	107	71	36
11		218	182	145	109	73	36

Speed.		*Shots on Target.*					
Secs.	*Fifths of a Second.*	**6**	**5**	**4**	**3**	**2**	**1**
10	4/5	222	185	148	111	74	37
10	3/5	226	189	151	113	75	38
10	2/5	231	192	154	115	77	38
10	1/5	235	196	157	118	78	39
10		240	200	160	120	80	40
9	4/5	245	204	163	122	82	41
9	3/5	250	208	167	125	83	42
9	2/5	255	213	170	128	85	43
9	1/5	261	217	174	130	87	43
9		267	222	178	133	89	44
8	4/5	273	227	182	136	91	45
8	3/5	279	233	186	140	93	47
8	2/5	286	238	190	143	95	48
8	1/5	293	244	195	146	98	49
8		300	250	200	150	100	50
7	4/5	308	256	205	154	103	51
7	3/5	316	263	211	158	105	53
7	2/5	324	270	216	162	108	54
7	1/5	333	278	222	167	111	56
7		343	286	229	171	114	57
6	4/5	353	294	235	176	118	59
6	3/5	364	303	242	182	121	61
6	2/5	375	313	250	188	125	63
6	1/5	387	323	258	194	129	65
6		400	333	267	200	133	67
5	4/5	414	345	276	207	138	69
5	3/5	429	357	286	214	143	71
5	2/5	444	370	296	222	148	74
5	1/5	462	385	308	231	154	77
5		480	400	320	240	160	80

Speed.		*Shots on Target.*					
Secs.	*Fifths of a Second.*	**6**	**5**	**4**	**3**	**2**	**1**
4	4/5	500	417	333	250	167	83
4	3/5	522	435	348	261	174	87
4	2/5	545	255	364	273	182	91
4	1/5	571	476	381	286	190	95
4		600	500	400	300	200	100
3	4/5	632	526	421	316	211	105
3	3/5	667	556	444	333	222	111
3	2/5	706	588	471	353	235	118
3	1/5	750	625	500	375	250	125
3		800	667	533	400	267	133
2	4/5	875	714	571	429	286	143
2	3/5	923	769	615	462	308	154
2	2/5	1,000	833	667	500	333	167
2	1/5	1,091	909	727	546	364	182
2		1,200	1,000	800	600	400	200

APPENDIX XII

PROOF AND ARMOURERS' MARKS

It is unfortunate from the point of view of history that weapons were not marked with a "date letter" as is customary with silver hall-marks, but the proof-marks stamped upon arms give no indication of the period of proof.

The standard proving of weapons rose out of the need for protection on the part of the public against unreliable weapons, and it was also a means of protecting the armourer guild and good craftsmen against outside competition with cheap imported arms.

Originally evolved by the armourers themselves, the system of proof was in nearly every arms-producing country taken over by the Municipality or the State, who imposed a series of stringent by-laws and penalties on any who sought to evade the prescribed tests. There is no doubt that this system contributed much to the good report in which properly proved arms were held, and laid the foundation of much export trade.

Proof marks are the official stamps or ciphers showing that a barrel has been properly tested with the prescribed excess charges of powder and ball that allow a large factor of safety to the user.

These marks are customarily the cipher of the gunsmiths' guild and the mark of the State or city.

Armourers' marks are quite distinct from proof-marks, and comprise the private marks, ciphers, stamps, or signatures of the individual armourers or gunsmiths. They are often found in combination with proof-marks, and still more frequently on unproved arms made in countries where no official proof was existent. Most specimen marks are of this nature, and it is no unusual thing to find the marks of two different smiths upon the same weapon, one man signing the lock work, the other being the barrel maker.

A great deal of painstaking research work is necessary before a list of armourers' marks applied to pistols can be compiled, but I have every hope of preparing such a list at a future date when peace is restored in Europe and the great museum collections are accessible. Besides these national collections, there are many municipal and private galleries whose curators have formed most excellent catalogues, but in the main this knowledge is unco-ordinated. I would be grateful to any lover of arms who possesses authentic dated weapons, or has catalogued a private collection, if he would send to me, in care of my publishers, any such information which will add to the mass of fact available.

The system of proof in Great Britain was as follows:

In England, the second charter of the Gunmakers' Company granted in 1672 marked the beginning of the official proving of arms, but it did not apply to weapons made elsewhere in the Kingdom, provided they were not offered for sale in the London market.

Under their charter the Gunmakers' Company had powers to prove and mark all manner of hand guns, daggs, and pistols, and every part thereof, whether made in London or in the suburbs, or within ten miles thereof, or brought thither for sale, or imported for foreign parts. To establish this rule, they constructed a proof house near the City of London.

Their cipher or mark was the letters G.P. interlaced, and surmounted by a crown. The viewer's mark was a capital V, also surmounted with a crown. These two, and sometimes the maker's private mark, may be found on many old weapons.

Birmingham had also a proof, but this test was not obligatory, and makers evaded the tests.

Birmingham did not establish its proof houses until 1813, and though another Act was passed in 1815 to strengthen the former, the laws were not stringent enough, and the Birmingham proof was not recognized as sound until much later, when the Act of 1855, still unsatisfactory, and the eventually satisfactory one of 1868, was passed.

In 1888 it was decided by a joint meeting of the Gunmakers' Company and the Birmingham Guardians that a new supplementary proof for smokeless powder should be instituted in 1893, 1896, and 1904; the rules were further adjusted to allow for new powders.

A voluminous schedule of rules and scales of charge for the proof of various kinds of small arms is in existence, and the proof is the same at London and at Birmingham, though different proof marks are employed. Modern pistols and revolvers are proved in their last stage of manufacture when complete with action and in firing order. Revolvers are marked on barrel, on the frame, and on each chamber.

Liège, in Belgium, has always been celebrated for its weapons, and established its proof house as a municipal enterprise on May 10, 1672, when it was laid down that barrels should be marked with the stamp of the city, and unproved weapons might not be sold under pain of a fine of a gold florin and the confiscation of the unproved barrel.

The laws were remodelled in 1849, and definite rules as to powder charges laid down, and the cipher, as shown on the right, granted as sign of definite proof.

In 1849 the additional ciphers for provisional proof the letters E. L. in script, enlaced, and the sign of the tower of Liège a little obelisk upon a pyramidal base, also shown on the right.

Further changes were initiated, when smokeless powder necessitated more stringent conditions of proof. These marks are the Belgian Lion rampant over the letters S.C.H. or E. C.

St. Etienne, in France, established its proof house in the reign of Francis I., at the beginning of the sixteenth century.

Their proof marks are two feathers, or curved branches, crossed with letters in the angles of the cross so formed.

The first cipher is the crossed branches surmounted by a crown, and with a cross or star in the three other angles. Underneath the whole appears the word St. Etienne. The second is the crossed branches, with the letters A.E. The third is a plain stamp, with the letters A.V.E. The fourth the crossed branches, with the letters S.E.E., and a cross in their angles. The fifth a crown, with three crossed spikes set over a capital E.

PLATE OF PROOF-MARKS.

ENGLISH PROOF-MARKS.

LONDON PROOF-MARKS.

View-mark.

Provisional Proof.

Definitive Proof of Barrel Proved once only.

Gunmakers' Company Proof-mark.

Nitro Powder Proof.

Example of Early London Proof-mark. Crown and G.P. Cipher is the Proof. Fleur-de-lis, the Maker's Mark. Crown and V, the Viewer's.

BIRMINGHAM PROOF-MARKS.

Proof-mark (old Pistols).

View-mark (old Pistols).

Provisional Proof.

Definitive Proof of Barrels Proved once.

Proof-mark (Present).

View-mark (Present).

Nitro Proof.

BELGIAN PROOF-MARKS.

Provisional.

Nitro Proof. The Lion over S.C.H. and E.C. is also used for Nitro Proofs.

Definitive.

Rifled Arms and Revolvers.

FRENCH PROOF-MARKS.

St. Etienne Proof.

St. Etienne Proof.

French Government Arms.

Imported Arms Proved in France.

St. Etienne Proof.

Paris Proof.

Paris Mark.

AUSTRO-HUNGARIAN PROOF-MARKS.

Definite Proof, Buda-Pesth.

Ferlach Proof.

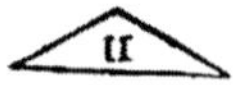

Double-barrelled Arms, Buda-Pesth.

Single-barrelled Arms, Buda-Pesth.

Wiepert Proof.

Vienna Proof

Prague Proof

GERMAN PROOF-MARKS.

German Proof.

Proof for Rifled Arms.

Smooth Bores.

German Double Proof.

Choke Bores.

Calibre Stamp.

SPANISH ARMOURERS' AND PROOF-MARKS.

Spanish Armourers' Stamps.

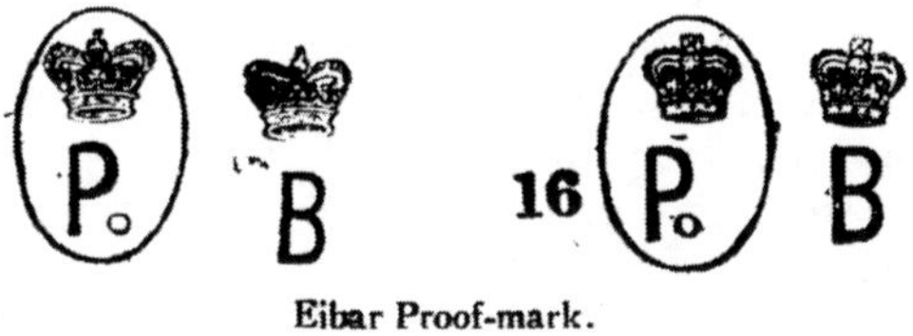

Eibar Proof-mark.

Other French towns had their special marks, such as—

Paris—a cipher of E. and P. set back to back (as shown on the right).

Sailleville—an E. reversed set in a square stamp (as shown on the right).

Chatellerault—a capital E.

Mutzig—a star above a capital E.

Tulle—a three spiked crown or mitre above a capital E (as shown on the right).

The Standard mark for French Government arms was a crown above an E., and this crown has a three looped top, with the orb or ball and cross as ornament, and is so distinguished from the cross-crested crown of some St. Etienne weapons.

Austrian arms were marked with the double eagle, and later with the letters Wn 78; a variety of private marks of manufacturers are also in use.

Spain depended for long upon the proof of individual armourers, and most Spanish barrels were marked with the stamp of the maker set in a square or oblong block under a crown. These marks have all the characteristic appearance of a rectangle, with a small blob at the top rather like a tied up sack of corn. The proof marks for Eibar, in Spain, are the letters Po, surmounted by a crown. In some cases this cipher is also preceded by the figures 16, and followed by a capital B., also surmounted by a crown.

U.S.A.—Most pistols exported from the United States are not proof marked in their country of origin. Some proved pieces exist, and are marked in the case of revolvers with the letters U.S., in the case of other arms with the letters V.P. and an eagle's head.

BILLING AND SONS, LTD., PRINTERS, GUILDFORD, ENGLAND

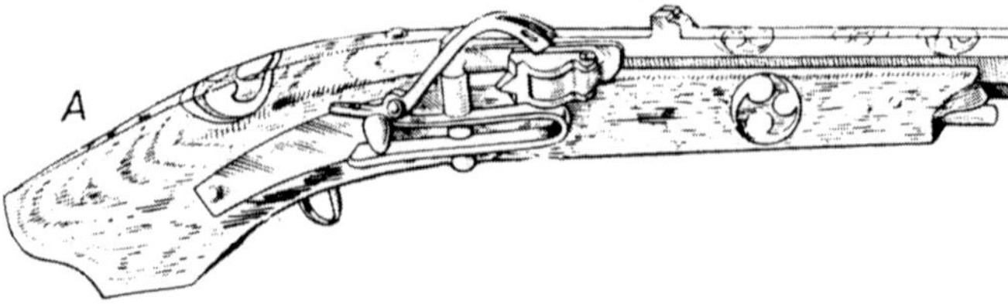

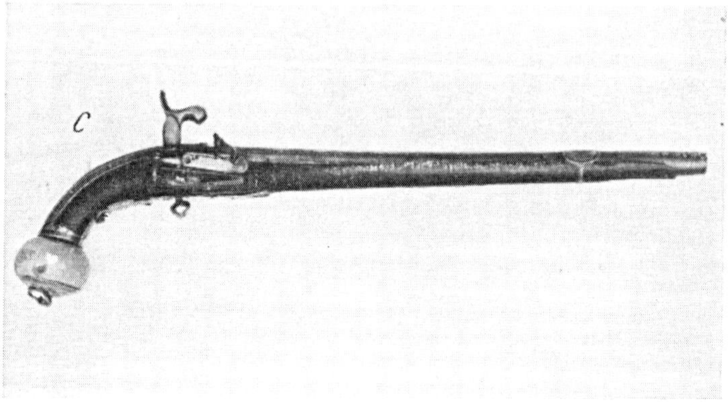

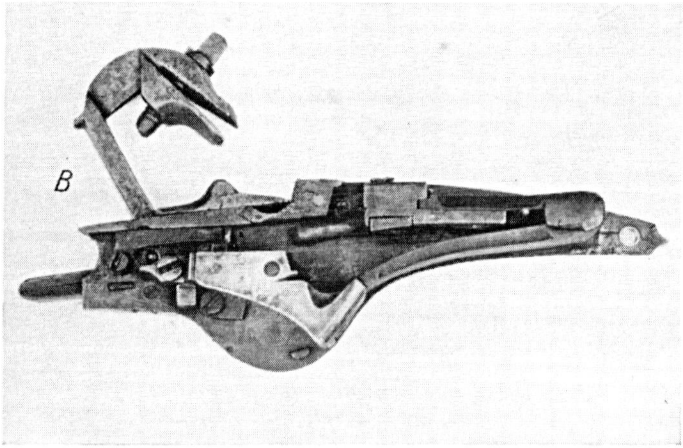

(a) Japanese Matchlock (circa 1542)

Copied by the Japanese from contemporary Dutch weapons

(b) Inside Mechanism of Wheelock

(c) A Russian or Circassian Pistol

Note the distinctive ball butt and ball trigger, also the Spanish type of lock. The mountings are of enamelled silver (Niello work) and the butt of ivory. The wood is all covered with fine grained leather. This specimen has been converted from flintlock to percussion

PLATE 2.

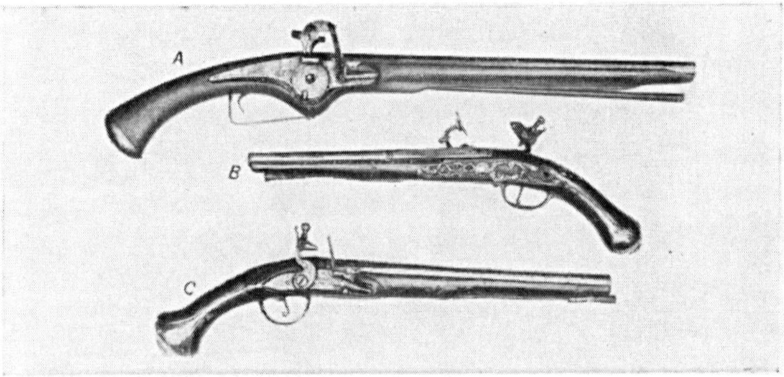

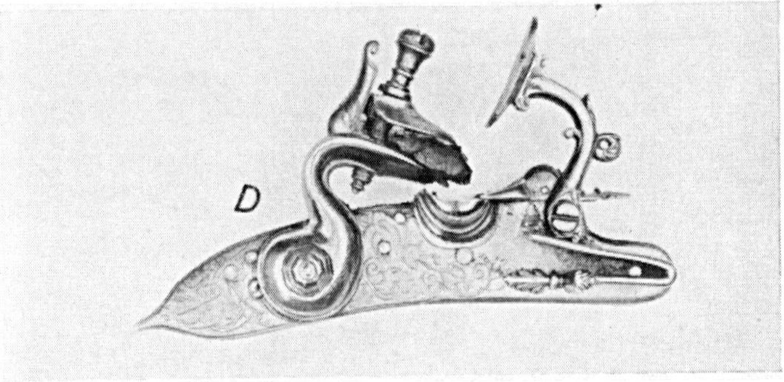

(a) Wheelock Petronel. Period Charles I. English

Note plain, serviceable appearance

(b) Italian Snaphaunce. Excellent workmanship and decoration.

Note how the pan cover and frizzen are independent. Period Charles I-II.

(c) Long-barrelled Italian Pistol. Period James II.

Note knob butt and florid decoration. Typical horse or holster pistol of the period

(d) Snaphaunce Lock

(e) Inside view of Snaphaunce Lock

PLATE 3.

Outside of Spanish and Oriental Lock with mainspring removed to show hammer extension and pin that serves as scear

Note fine engraving of steel

Inside of Spanish or Oriental Lock showing pivoted scear pin and spring

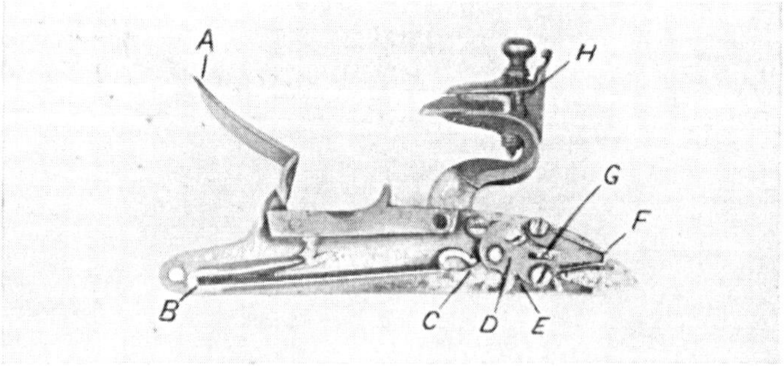

PARTS OF FLINTLOCK

(a) Frizzen Pan Cover or Steel. (b) Mainspring. (c) Tumbler. (d) Bridle. (e) Scear. (f) Scear-spring. (g) Safety Catch. (h) Hammer or Cock

PLATE 4.

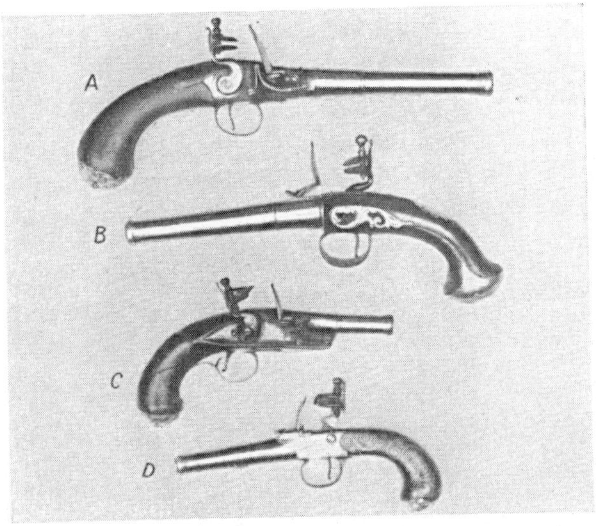

(a) Queen Anne Cannon Barrel Pistol

Note frizzen spring round the pan. Silver mask on butt dated 1705

(b) Queen Anne Cannon Barrel

With "fishtail" butt, silver escutcheons

(c) Queen Anne Pocket Cannon Barrel by Bunney

Note wood forend, rounded hammer forging, and masked butt

(d) Pocket Cannon Barrel, but later type, not Queen Anne

Note top action and enclosed lock and ring hammer. Date of hall mark on butt, 1825. Stock inlaid wire

1 2 3 4 5

Types of Hammers

(1) Early round section hammer hand forged
(2) Later flat section, stamped from grey steel
(3) Ring or double necked hammer
(4) French hammer with under spur
(5) Percussion

PLATE 5.

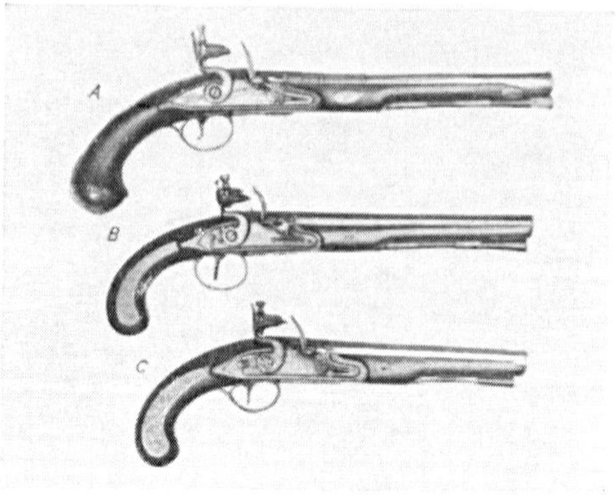

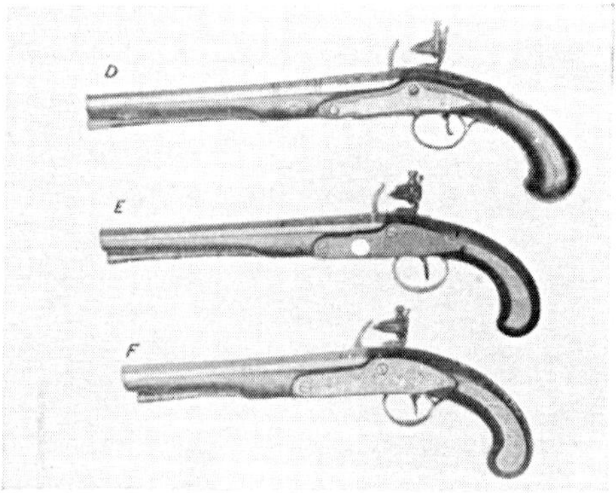

(a) An Irish Duelling Pistol

Note the ball stock, the round swamped barrel and the lock of later date than the stock and barrel

(b) A Standard English Flintlock Duelling Pistol

(c) A Brass Barrelled Military Pistol of Military Bore, but built on duelling lines

Note silver inlay and gilt gunmetal mountings

Left-hand Side View of same Pistols

(d) Note Old Engraved Lock Strap

(e) Plain but for Silver Escutcheon. Typical duelling type, 1790

(f) Details of Inlay show Squirrels eating Nuts.

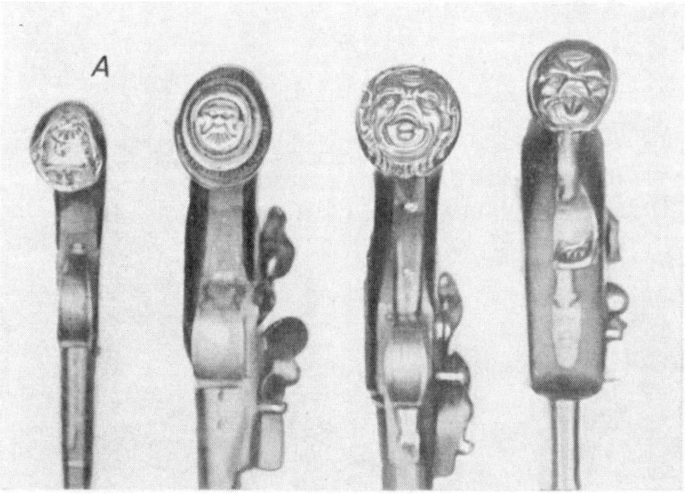

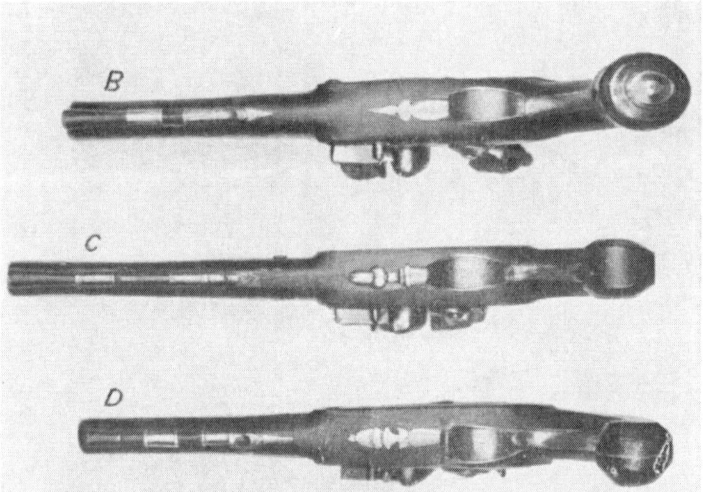

(a) Silver Masks on Queen Anne Type of Pistol

CHANGES IN TRIGGER GUARD ORNAMENT ILLUSTRATED

(b) Narrow Ornament, 1730-50

(c) Acorn, 1745-80

(d) Pineapple, 1780-1880

PLATE 7.

PLATE 8.

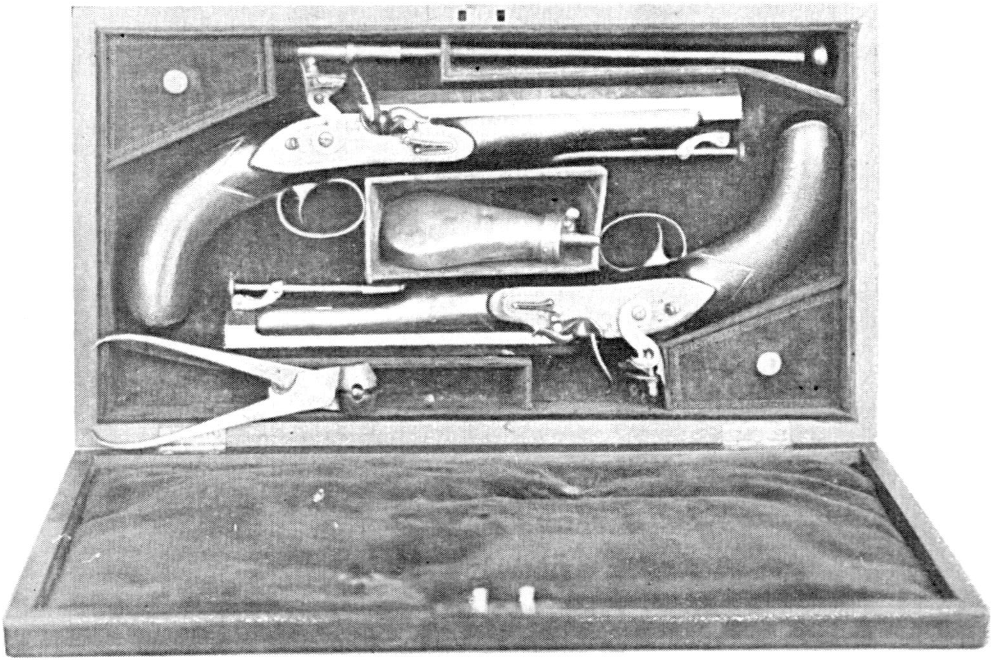

Case of Heavy Military Flintlocks by Scudamore. Period of Waterloo, 1815

Note small ring hammers, stirrup ramrods and heavy octagon barrels. The bore is that of the Brown Bess Musket

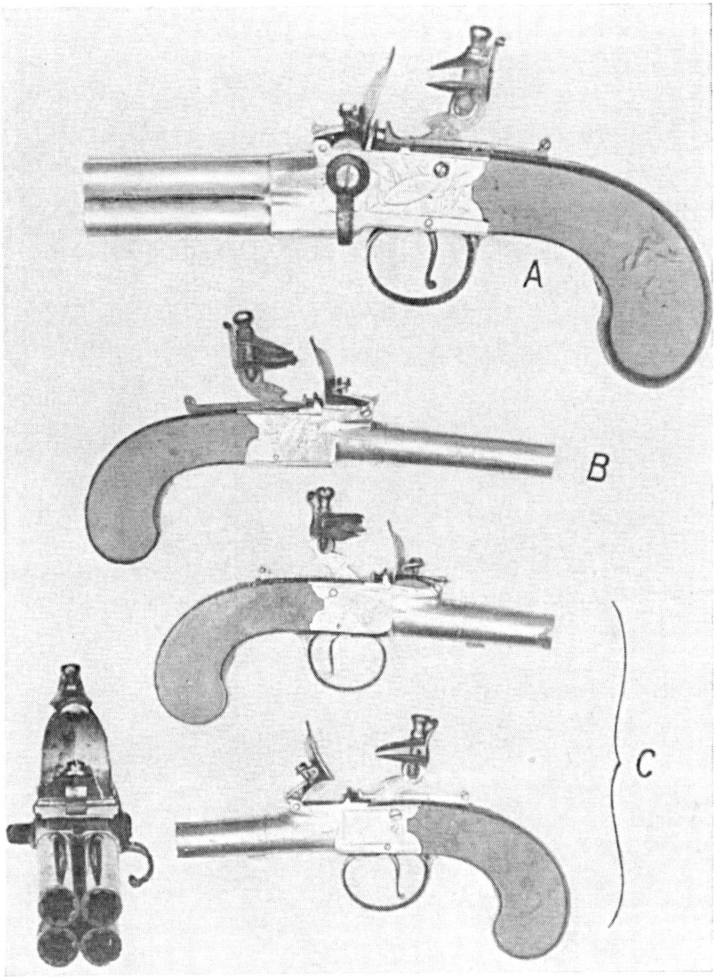

(a) A Four-barrelled Brass Repeating Flintlock by Knubley

Note the tap action by which successive priming charges are uncovered

(b) Common Pocket Pistol with Secret Trigger

(c) Pair of Small Pocket Pistols. These are typical pieces of the Standard Pocket Pistol from 1780 to 1850

Note the ring hammers and safety bolt to pan cover

(Inset) Four-barrelled Flintlock Repeating Pistol, front view of barrels showing coarse grooves for unscrewing barrel and side slide and tap action

PLATE 9.

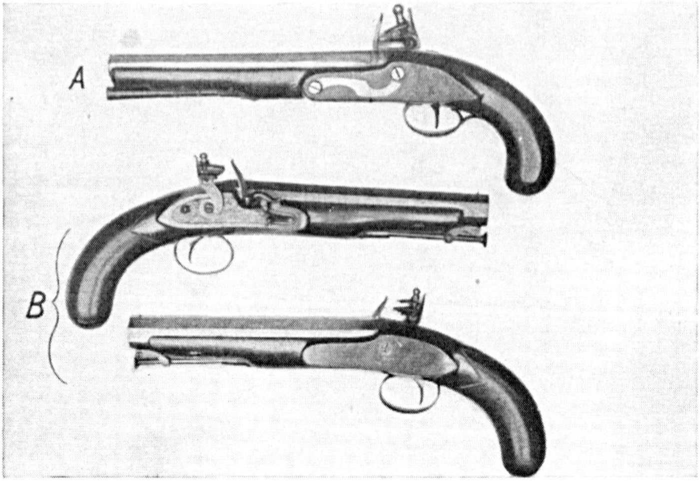

(a) Military Pistol. George II.

Note clumsy round barrel and brass lock strap

(b) Pair of Late Military Flintlock Pistols (Waterloo period) with small ring hammers and stirrup ramrods

Note side nail

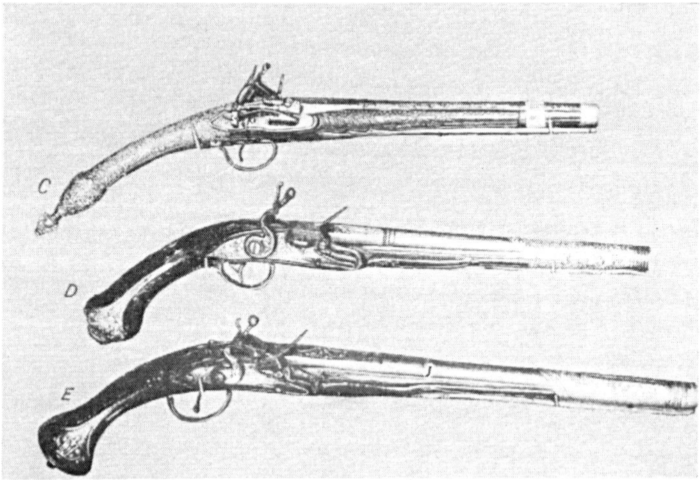

TYPES OF ORIENTAL WEAPONS

(c) Albanian or Circassian All-Metal Belt Pistol. Russian Lock

(d) Decorative Central European or Levantine Rubbish

(e) Similar piece

PLATE 10.

PLATE 11.

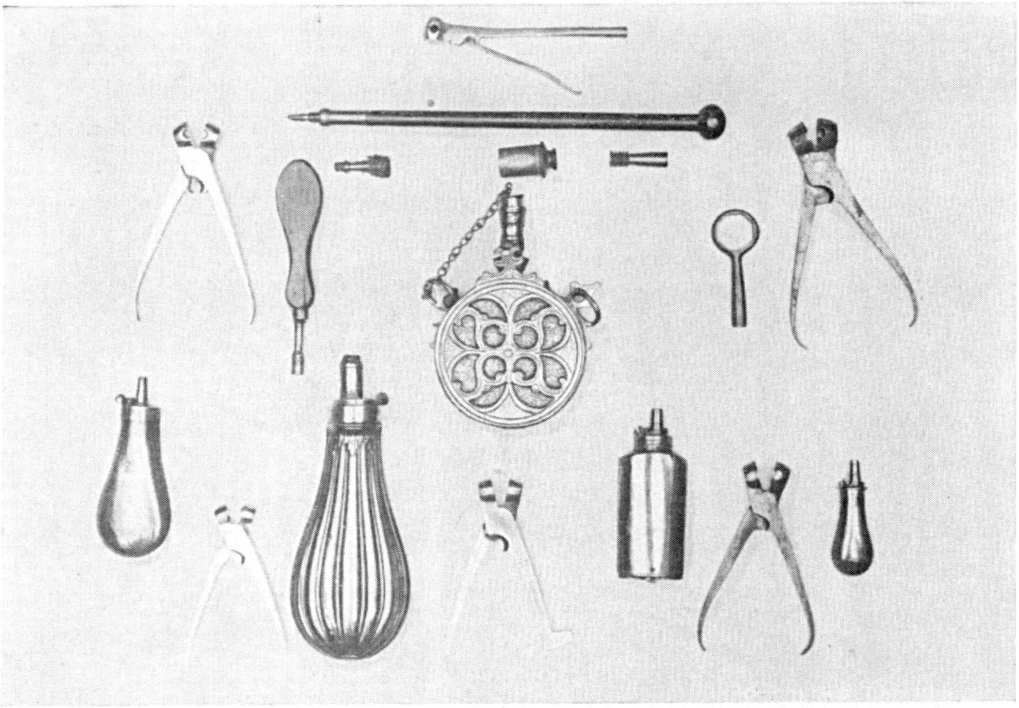

Accessories: Powder Flasks, Priming Flasks, Bullet Moulds, Cleaning Rods, Nipple Keys, Barrel Key and Oil Bottle

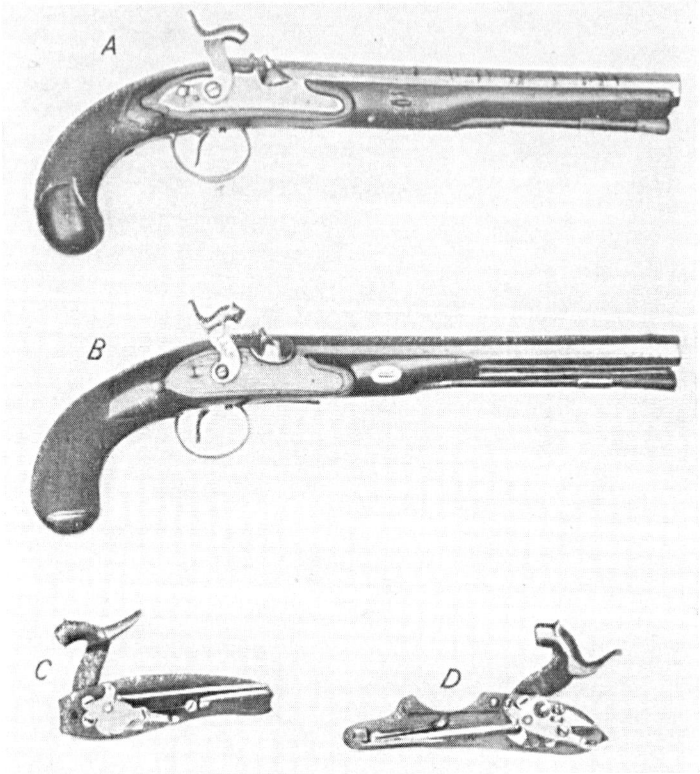

(a) Hair Trigger Percussion Duelling Pistols

Note wood stock continues right up to the end of the barrel

(b) A Pistol by Nock. The forend only reaches halfway up the barrel, and is finished with a horn tip. Ramrod is held to a lower rib brazed to barrel.

(In each pistol the hair trigger set screw is seen within the trigger guard in front of the trigger)

(c) Percussion Back Action Lock. 1850

(d) Percussion Lock. 1825

Note the "detent" in the notch of the tumbler

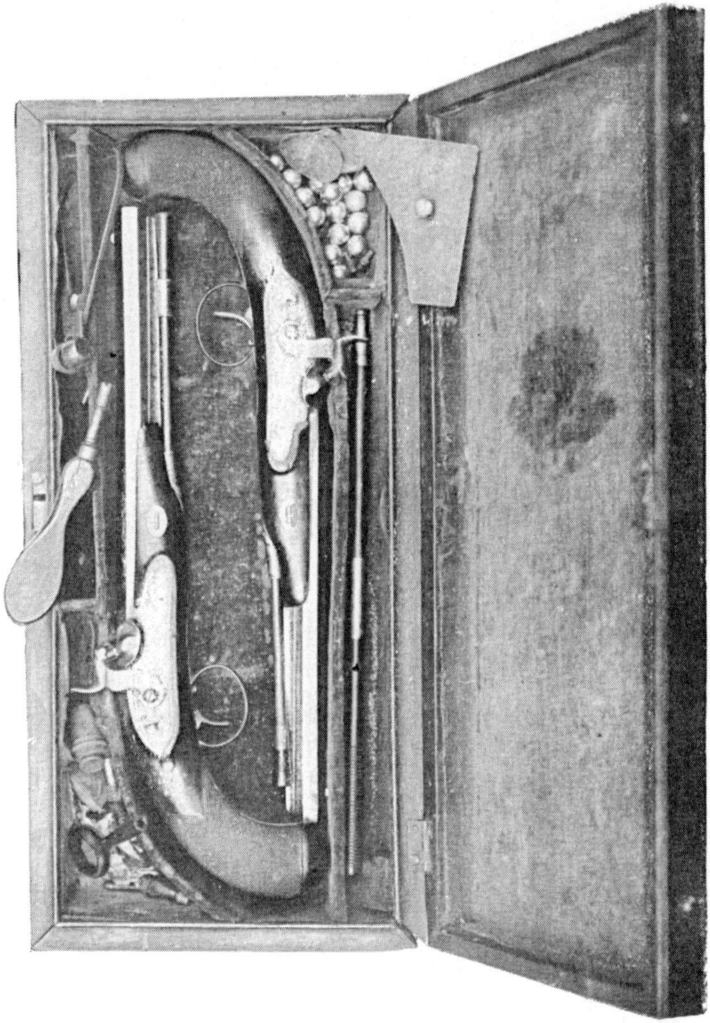

Case of Nock's Duelling Pistols with Accessories. 825

PLATE 13.

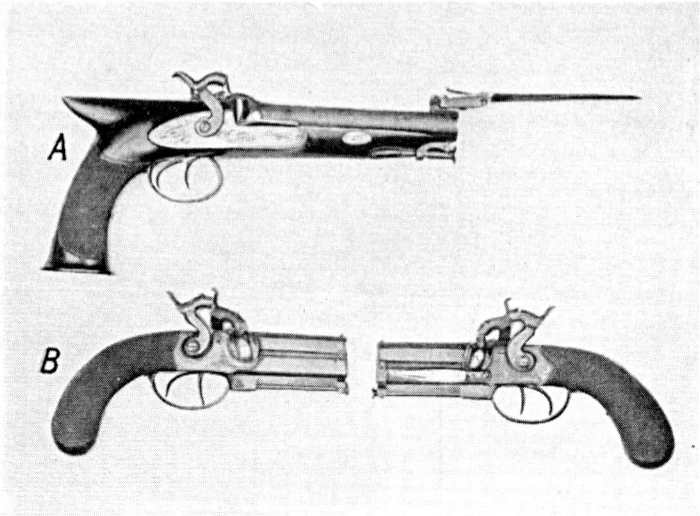

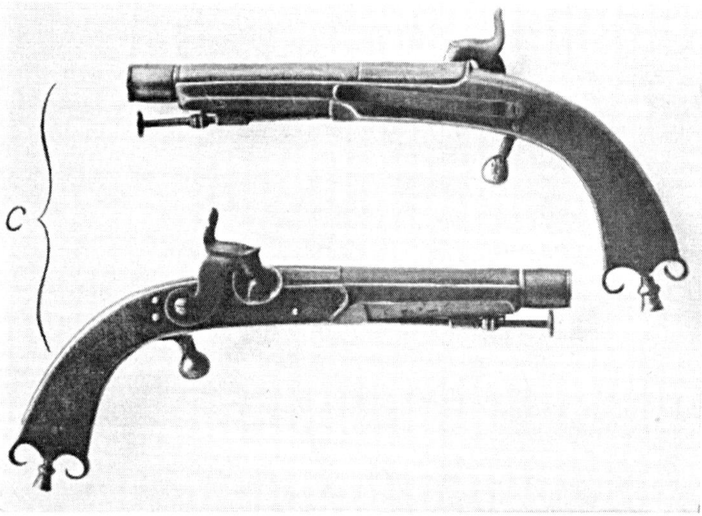

(a) Saw-handled Double-barrelled Rifled Percussion Pistol by Maybury.

Note folding spring bayonet

(b) Double-barrelled "under and over" Belt Fistols

(c) Scotch Percussion Pistols, about 1840, in possession of Mr. George Evans, of Wolverhampton

(They are a typical pair of late Scotch dress pistols)

PLATE 14.

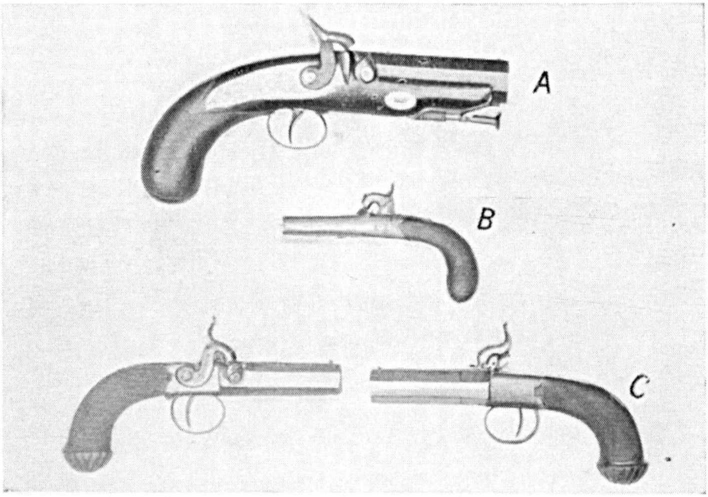

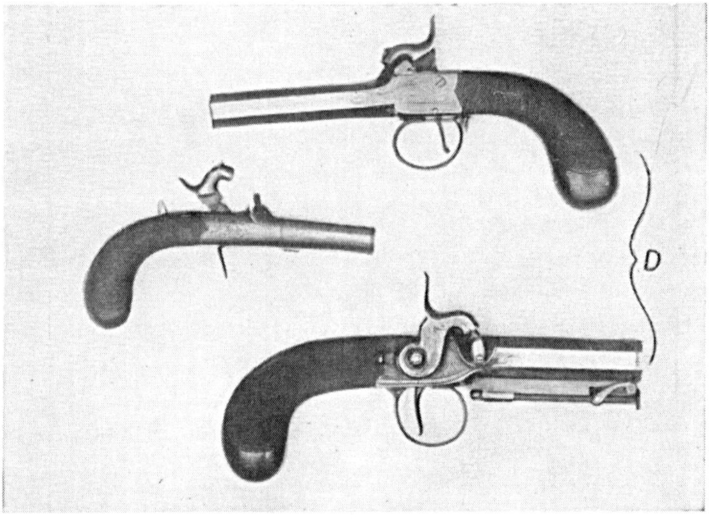

(a) Back Action Pistol by Egg. 12 bore

(b) Pocket Pistol with Folding Trigger by Powell

(c) Pair of Percussion Pocket Pistols by Burnand. Large bore, small action. Side Hammers

(d) Common Percussion Pocket Pistol

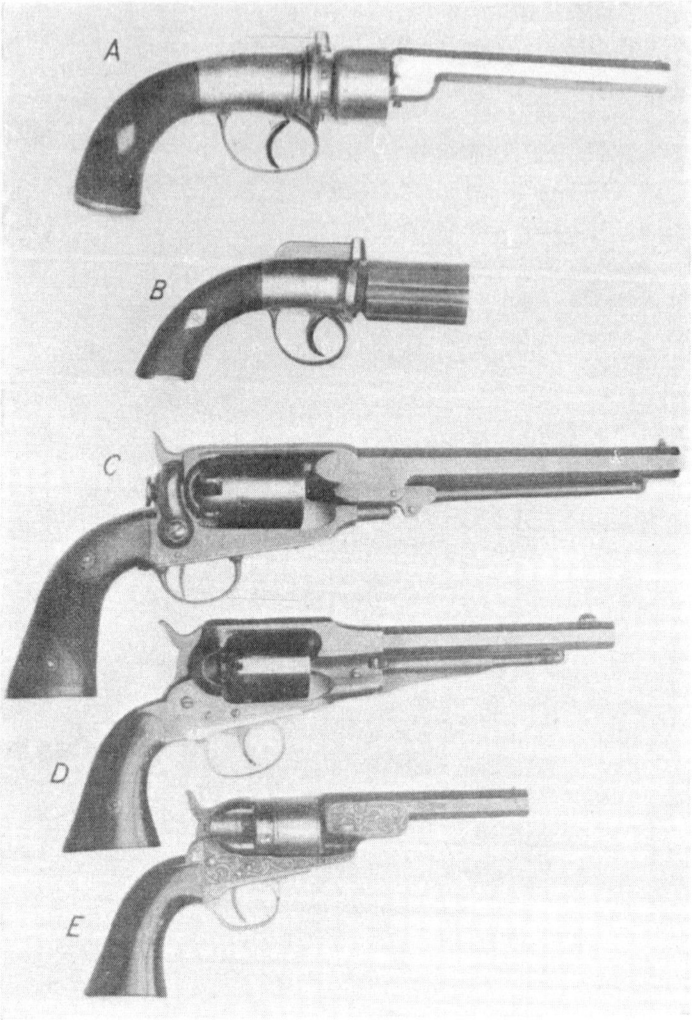

(a) Percussion Revolver, 1850 (c) Colt Side Hammer Revolver
(b) Pepperbox Revolver 1850 (d) Remington Revolver
(e) Small Colt Revolver, London

PLATE 16.

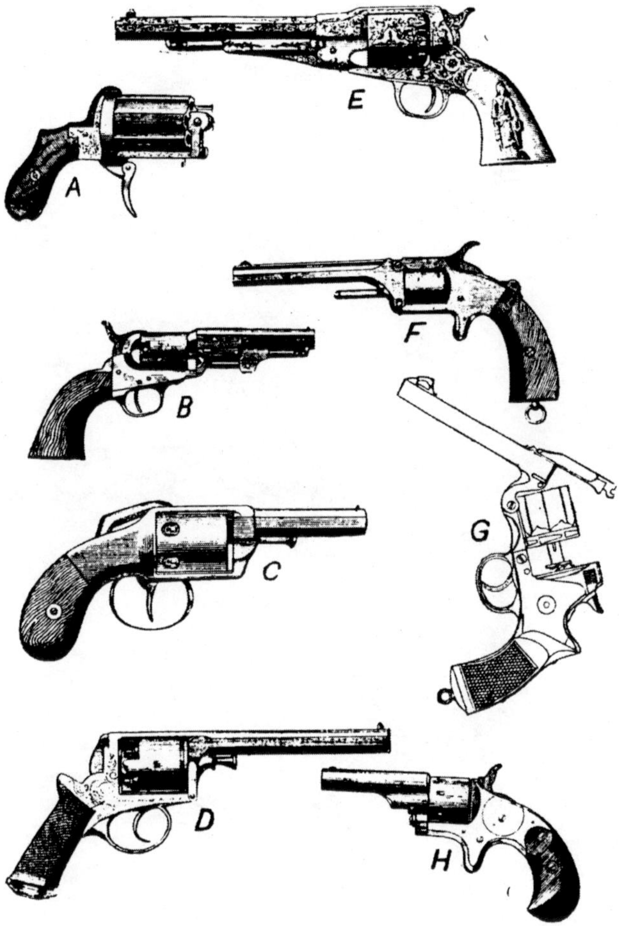

EARLY REVOLVERS

(a) Pinfire Revolver

(b) Colt (American)

(c) Forehand and Wadsworth (American)

(d) Adams (British)

(e) Finely Engraved Remington (American)

(f) Smith and Wesson (American)

(g) Enfield Service Revolver, '442 (British)

(h) Colt '22 Revolver (American)

PLATE 17.

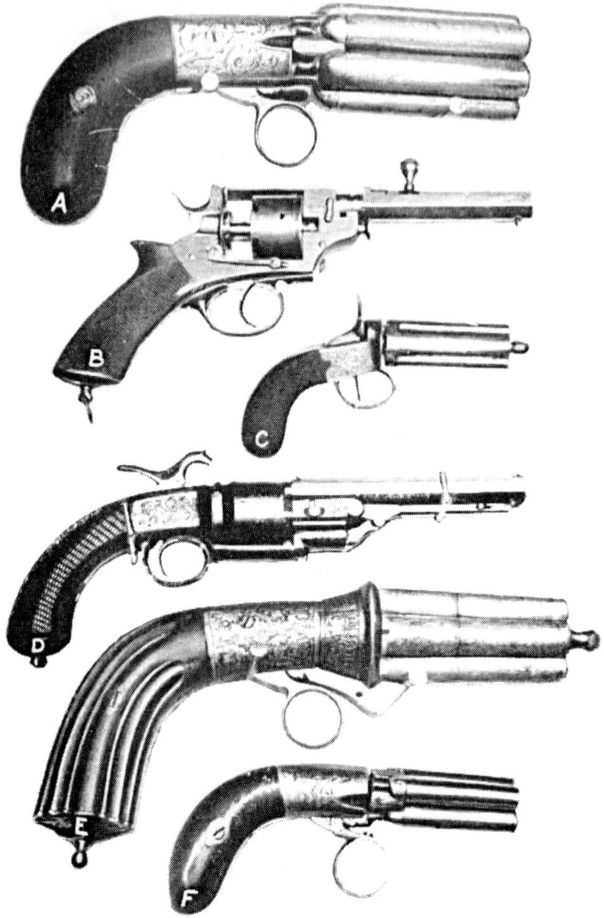

EARLY REVOLVER MECHANISMS

(a) Underneath Hammer, Belgian Comblain Model Muzzle Loader

(b) Thompson's Bolt Ejector Model

(c) French Rimfire 5-Shot Pepperbox

(d) Lang's Muzzle-loading Revolver

(e) Heavily Chased Pinfire Pepperbox by Lefaucheux, of Paris

(f) Belgian Muzzle-loading Pepperbox with separate Barrels screwed to Multiple Breech

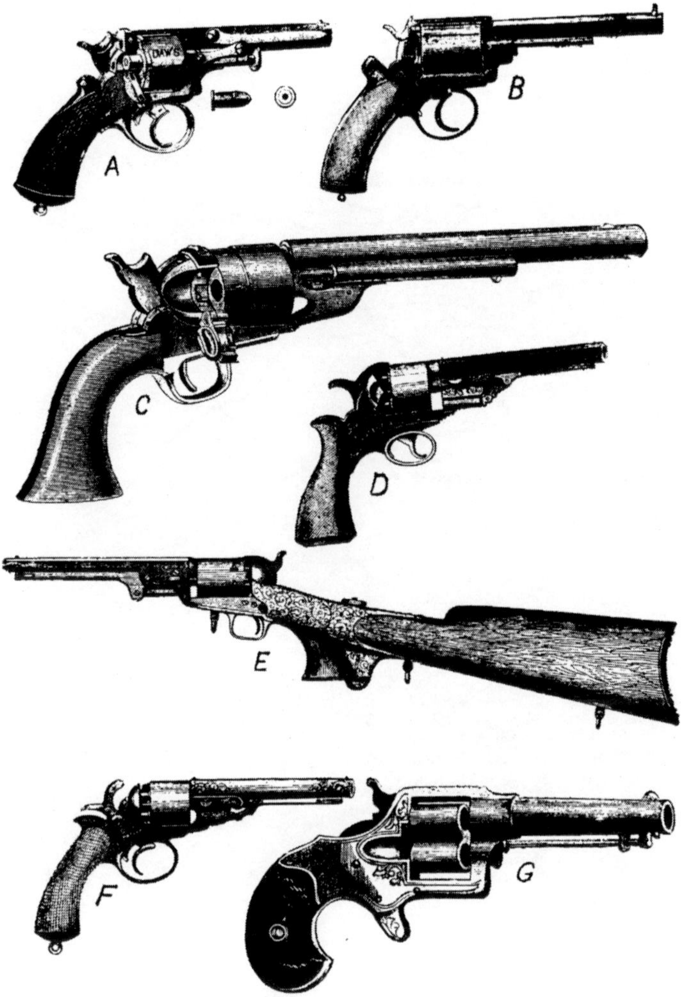

(a) Daw's Patent Revolver

(b) Tranter's Patent Revolver

(c) A Muzzle-loading Colt converted to Breech-loader

(d) Deane and Adam's Patent

(e) A Four-chambered Colt House Pistol

(f) A Bland Muzzle-loader Revolver

(g) A Colt Revolver fitted with Carbine Stock. London Navy Model

PLATE 19.

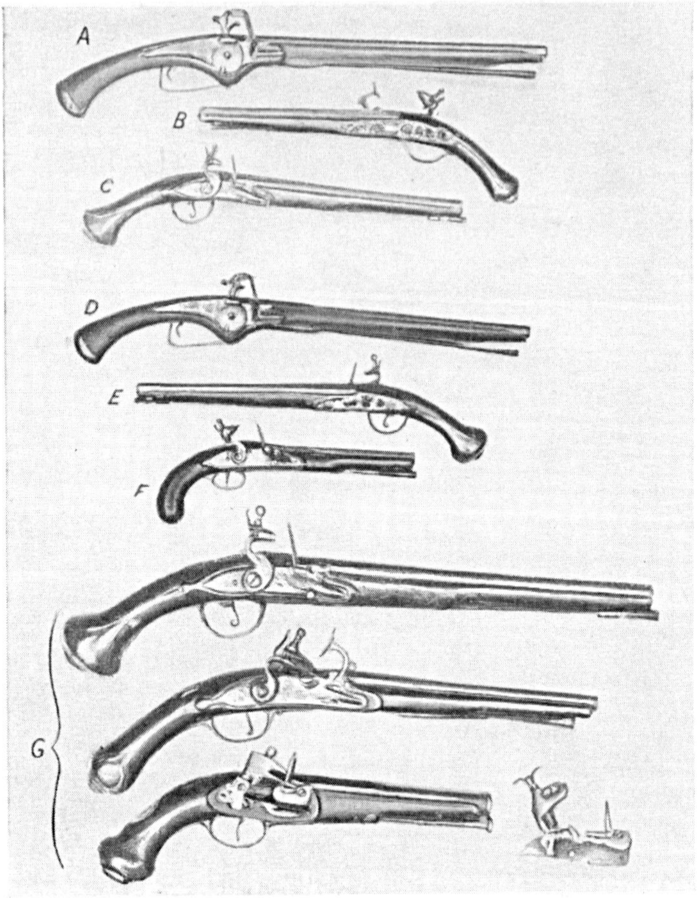

SERVICE WEAPONS.

(a) Wheelock, Elizabeth to Charles I.

(b) Snaphaunce, Charles II. to James II.

(c) Long Flintlock, 1700-60

(d) Wheelock, 1600-1700

(e) Long Flintlock 1700-50

(f) Short Flintlock, 1750-1800

(g) A Catalan Dag or Spanish Pistol

Note the typical Spanish lock, cannon barrel and external lock mechanism similar to Russian and Oriental flintlocks. This type of lock is sometimes miscalled by dealers as "snaphaunce." Contrast with B the real snaphaunce

PLATE 20.

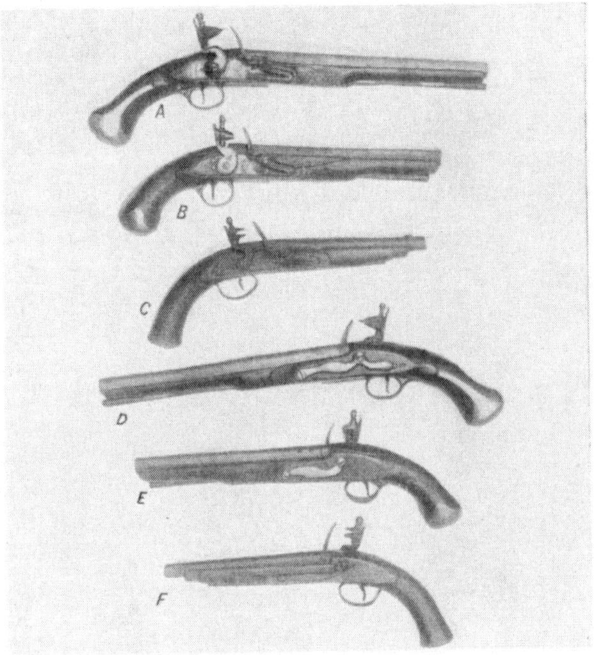

MILITARY PISTOLS

(a) Long Flintlock 20 Bore, dated 1759, Crown and G.R. on lock plate

Note swanneck hammer and barrel 12 in. long. Government weapon

(b) Late Military Pistol dated 1803

Note ring on double hammer and handier length of barrel, 9 in. Typical of Napoleonic period and in use until the Crimean War. Government weapon

(c) An Export Trade Pistol by Ketland of Birmingham

Inferior to Government arms, and as supplied to colonists of America and armed merchantmen. Typical ship's pistol of period 1790-1820

(d) Reverse Side of Plate.

Note distinctive lock-strap in brass scroll work, also scrolled tip to trigger

(e) Later period with less ornament about lock-strap and trigger guard

(f) Plain finish and with rough steel belt hanger for hanging weapon on sash or belt

Note how the brass butt knob of the Government weapons is here replaced by a simple brass plate

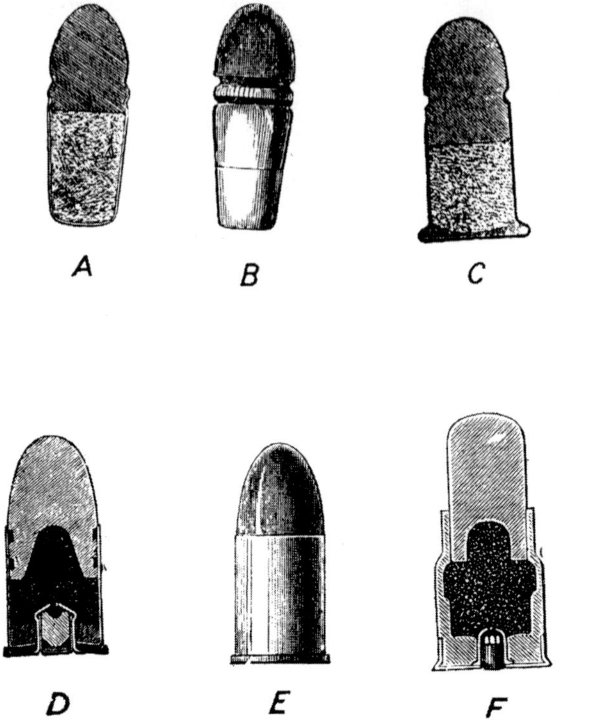

Early Cartridges for Revolvers

(a) Section of Muzzle-loader. (b) Outside view of same. (c) Section of Rim Fire. (d) Section of Centre Fire. (e) Outside of Centre Fire. (f) Section of Daw Pistol Cartridge

PLATE 22.

Standard Range of Revolver Cartridges—I.

PLATE 23.

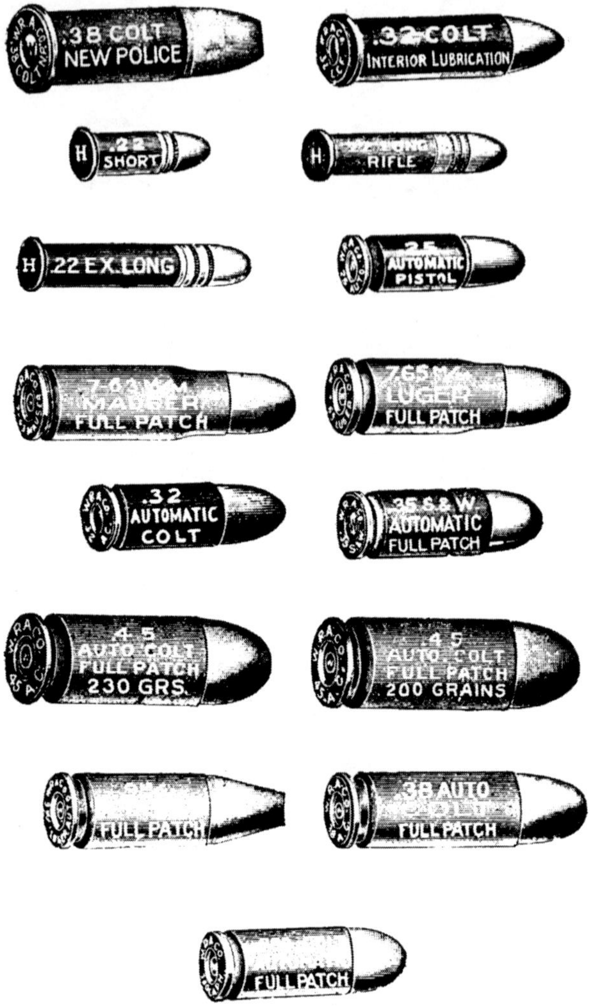

Standard Range of Revolver Cartridges—II.

Standard Range of Revolver Cartridges—III.

PLATE 25.

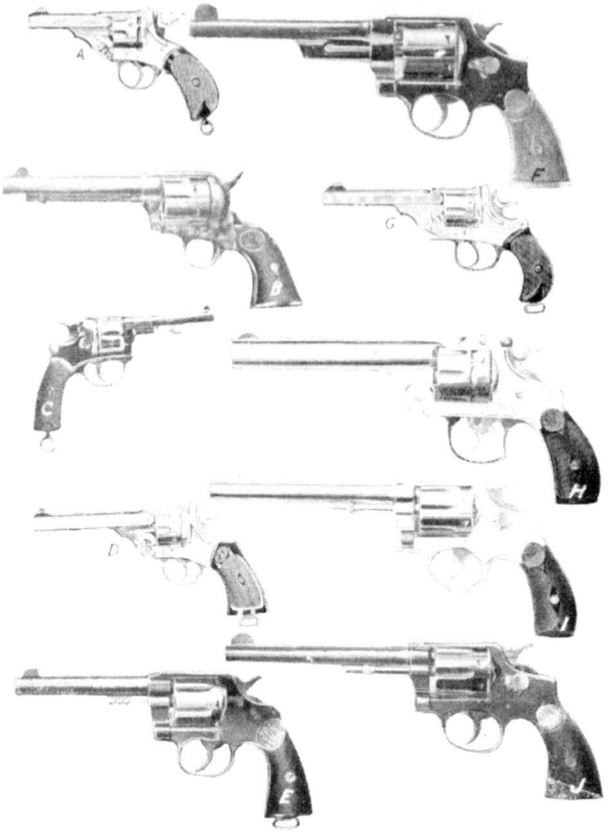

MILITARY REVOLVERS

(a) Webley Mark V. ·455

(b) Colt S.A. Army ·455

(c) French Ordnance Model ·38

(d) Webley Scott ·455

(e) Colt New Service ·455

(f) Smith and Wesson ·455

(g) Webley Army

(h) Smith and Wesson ·44 Russian (breakdown model)

Smith and Wesson Military Model 38 calibre, 1902

(j) Smith and Wesson Military Model 38 calibre, 1905

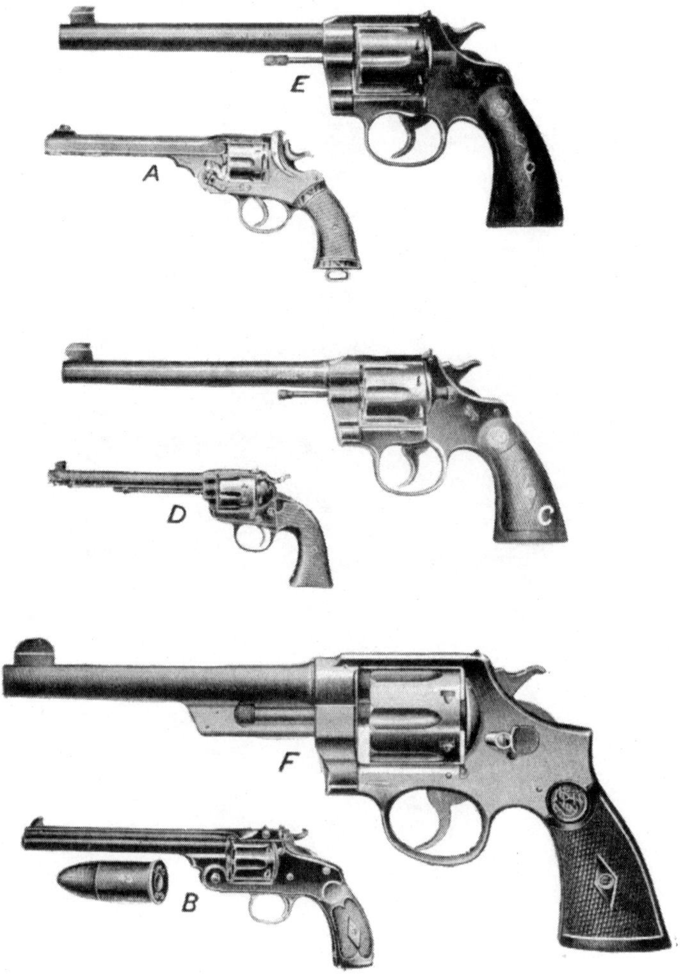

TARGET REVOLVERS

(a) W. G. Webley Target ·**455**

(b) Smith and Wesson S.A. Target ·**455**

(c) Colt Officers' Model ·**38**

(d) Colt Bisley Model ·**455**

(e) Colt New Service Target ·**455**

(f) Smith and Wesson Target ·**455**

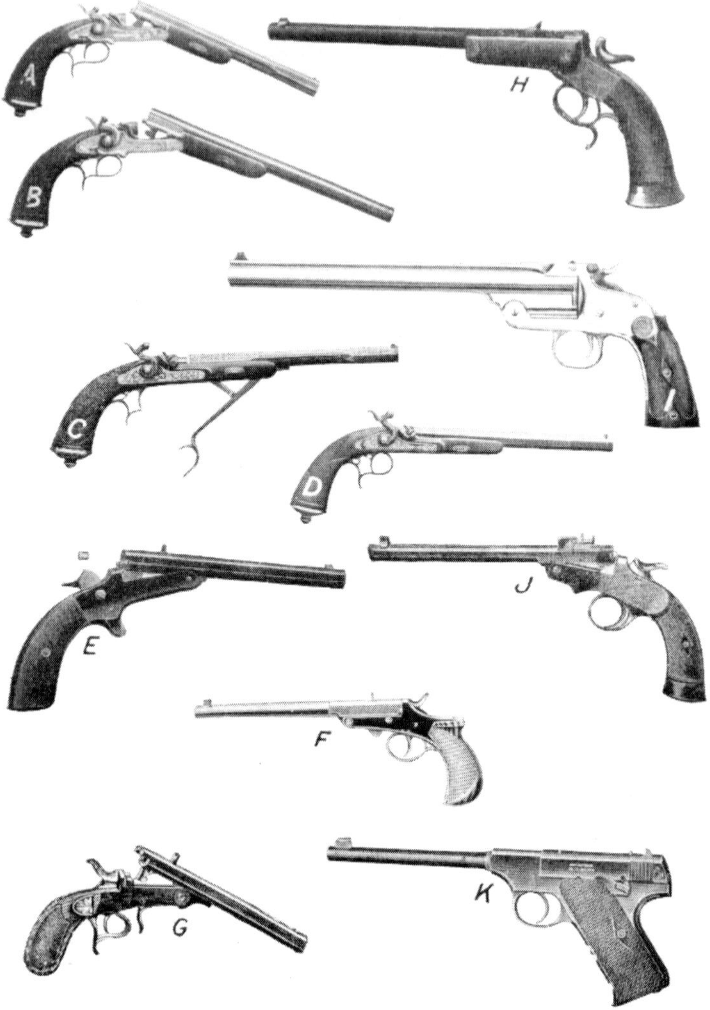

TARGET PISTOLS

(a) and (b) French ·44 calibre Pistols, made on modern duelling pistol lines

(c) A Gastinne-Rennette Semi-muzzle Loader for Duelling Practice

(d) Regulation Duelling Pistol

(j) Continental Target Pistol

(h) Stevens Target Pistol

(g) Continental "Saloon" Pistol

(i) Smith and Wesson Target Pistol

(f) Webley Target Pistol

(e) Continental Target Pistol

(k) Colt Automatic Pistol, Calibre ·22 Target Model (Latest)

PLATE 28.

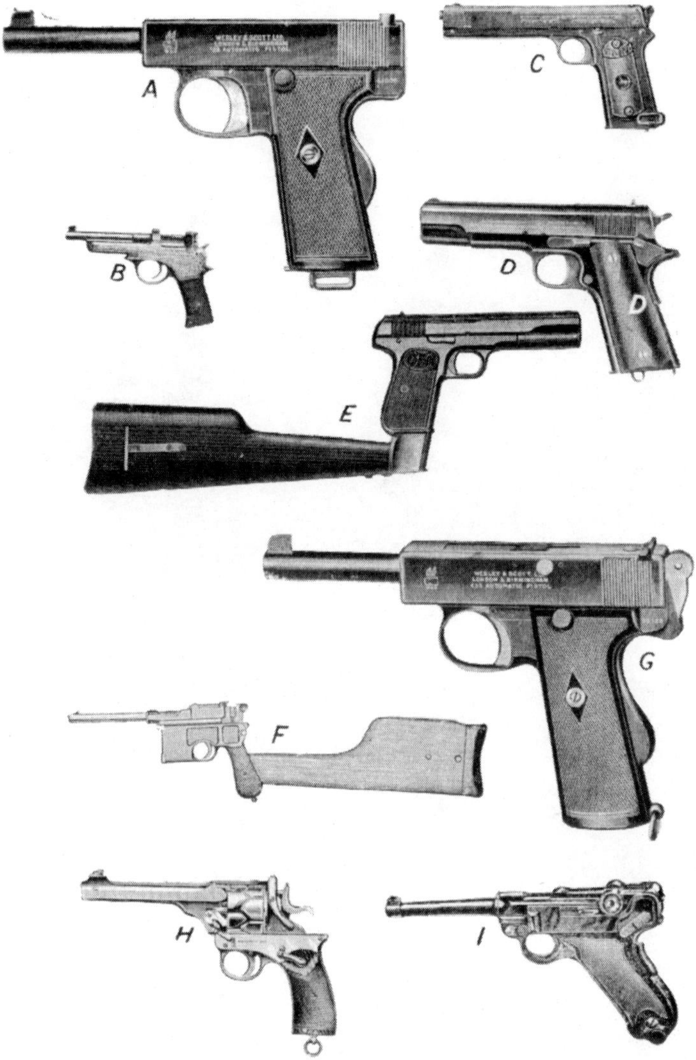

MILITARY AUTOMATICS

(a) Webley and Scott ·38
(b) Mannlicher 9 mm
(c) Colt ·38
(d) Colt ·45
(e) Browning 9 mm. long
(f) Mauser ·35
(g) Webley ·455 Auto
(h) Webley Fosberry ·455
(i) Luger Parabellum 9 mm ·380

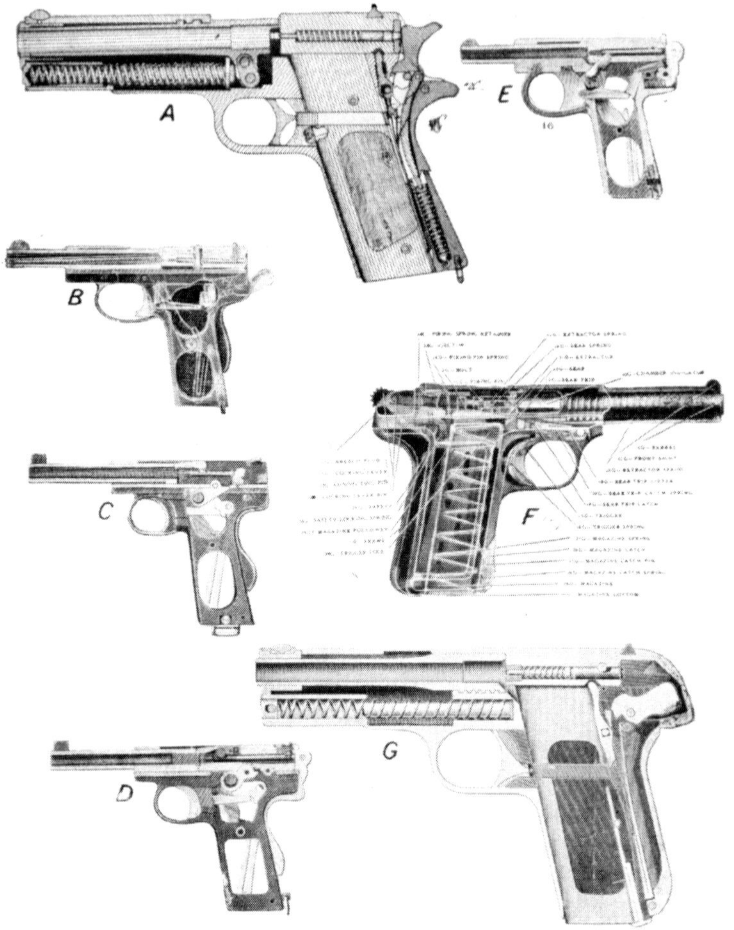

SECTIONS OF AUTOMATICS

(a) Colt ·45
(b) Webley ·455
(c) Webley ·38
(d) Webley ·380
(e) Webley ·32
(f) Savage ·32
(g) Colt ·32

PLATE 30.

POCKET REVOLVERS

(a) Colt Police Positive ·32-38

(b) Velo Dog ·25

(c) Smith and Wesson Hammerless

(d) Velo Dog ·38

(e) Smith and Wesson Pocket Models

(f) Smith and Wesson Perfected

(g) Colt Pocket Positive

(h) American Pocket ·32

(i) Webley Hammerless ·320

PLATE 31.

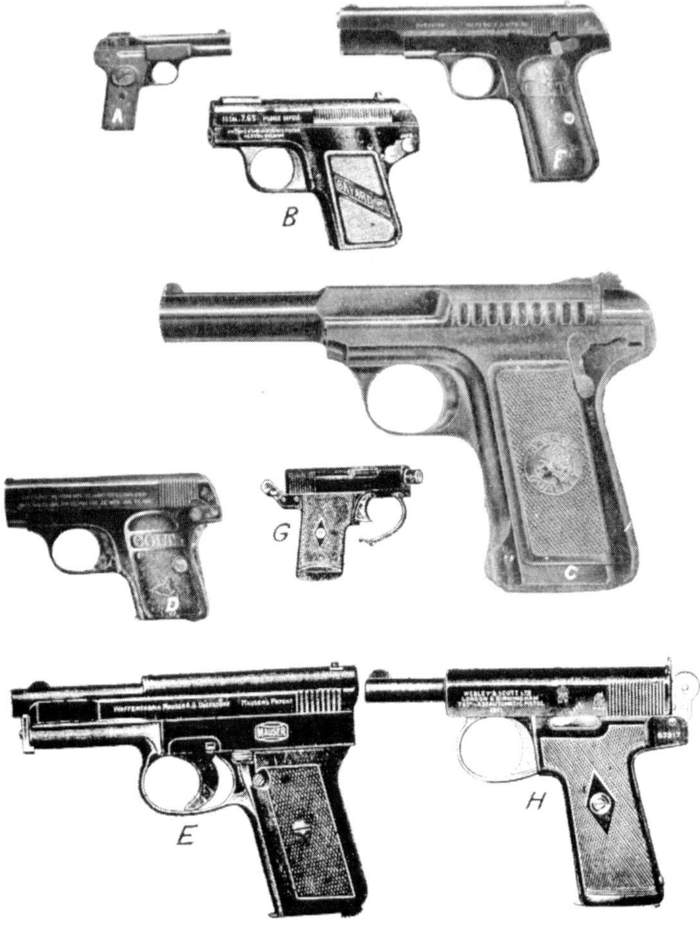

POCKET AUTOMATICS

(a) Browning Old Model ·32

(b) Bayard ·32, ·380, ·25

(c) Savage ·32, ·380

(d) Colt ·25

(e) Mauser ·25

(f) Colt ·32

(g) Webley ·25

(h) Webley ·320, ·380

PLATE 32.

POCKET PISTOLS

(a) Continental Derringer or Bicycle Pistol ·22 calibre

(b) Colt Derringer No. 1

(c) Remington Double Barrel ·41 calibre Derringer

(d) Ashland's Patent ·22 calibre

(e) "Reform" Model ·25 calibre

(f) Muzzle-loading Derringer

(g) Colt No. 2 Derringer ·41 calibre

(h) Remington Repeating Derringer ·32 calibre

(i) Colt Side Action Derringer ·41 calibre

(j) Double-barrelled Large Bore Saddle Pistols

PLATE 33.

(a) The Wrong Way to Grip a Revolver

(b) The Right Way to Grip a Revolver

(c) Standard Target Firing Position

(d) The Wrong Way to Grip an Automatic

(e) The Right Way to Grip an Automatic

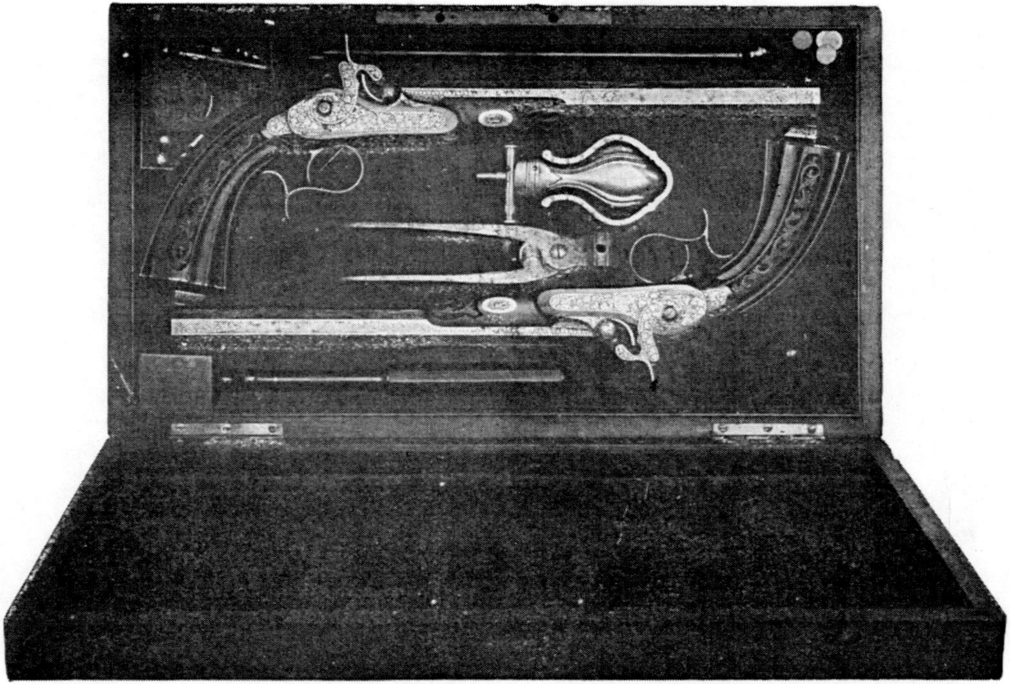

A Pair of Modern French Duelling Pistols by le Page Moutier

Note exquisite chiselled relief engraving, ebony stocks, and long, heavy barrels; also mallet and loading rod in case

PLATE 35.

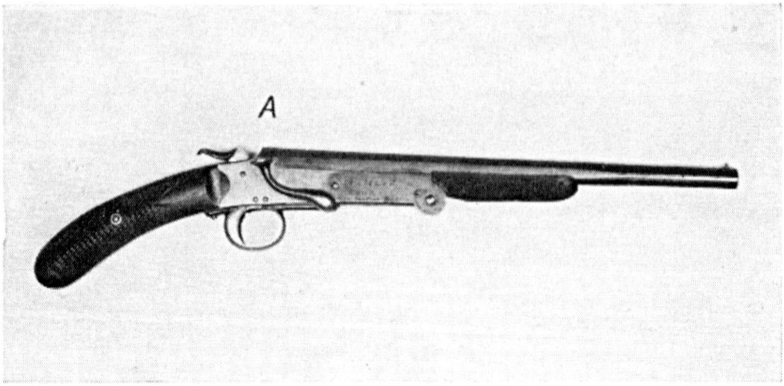

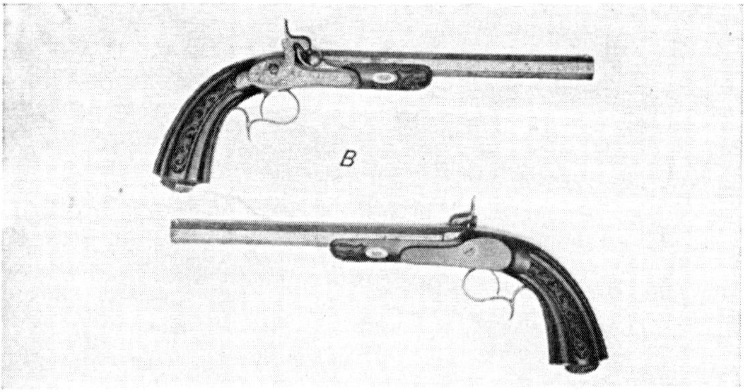

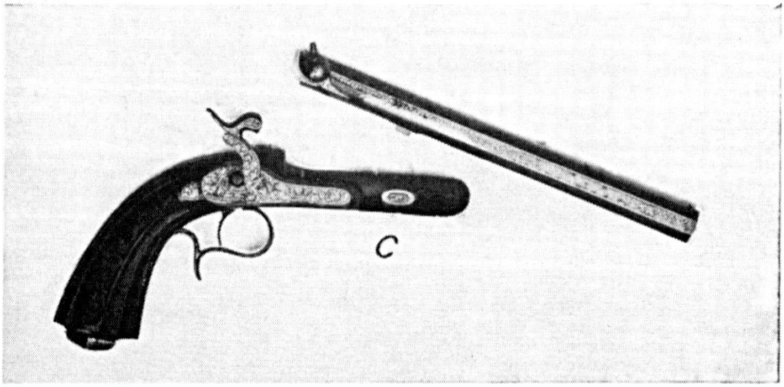

(a) A ·410 Bore Shot Gun Pistol

Although without a back sight, this weapon will make a 2 in. group with round bullets at 20 yards

(b) A Pair of Modern Standard Duelling Pistols (Pistolets de Tir) by Le Page, of Paris

(c) French Duelling Pistol with Barrel Dismounted from Stock for Cleaning

PLATE 36.

Firing Booths at Gastine Rennette's. The most famous pistol gallery in the world

PLATE 37.

Sportsman's Vintage Press

Available at www.SportsmansVintagePress.com

Sixguns..Elmer Keith
Sixgun Cartridges and Loads....................................Elmer Keith
Big Game Rifles and Cartridges................................Elmer Keith
Shotguns by Keith..Elmer Keith
Elmer Keith's Big Game Hunting.............................Elmer Keith
Keith's Rifles for Large Game..................................Elmer Keith
The Secrets of Double Action Shooting...................Bob Nichols
Wilderness Hunting and Wildcraft............................Townsend Whelen
Why Not Load Your Own..Townsend Whelen
.22 Caliber Rifle Shooting...C. S. Landis
Rifle-Craft..C. S. Landis
The American Deer Hunter..Francis Sell
Asian Jungle, African Bush.......................................Charles Askins
Modern Shotguns and Loads.....................................Charles Askins
The African Hunt..Charles Askins
African Rifles and Cartridges...........................John "Pondoro" Taylor
Manual of Police Revolver Instruction......................R. M. Blair
The Elusive Ten...William Reichenbach

Smith & Wesson Hand Guns....................................McHenry & Roper
Hunting Pronghorn Antelope...................................Bert Popowski
The Woodchuck Hunter...Paul Estey
Small Bore Rifle Shooting......................................Edward Crossman
Modern Bow Hunting...Hiram Grogan
Salt Water Fishing Is Easy.......................................Jerry Sylvester
Fishing Flies and Fly Tying.....................................William Blades
This Business of Exploring.............................Roy Chapman Andrews
Handbook for Shooters and Reloaders......................P. O. Ackley
Greatest Fishing..Joe Brooks
Hunting Our Biggest Game.......................................Clyde Ormond
How To Hunt Deer..Edward Freeman
Burning Powder...Douglas B. Wesson
Practical Rifle Shooting..Walter Winans
Handloading Ammunition...J. R. Mattern
Shooting..J. Henry Fitzgerald
Rifles & Rifle Shooting..Charles Askins
Mauser Rifles and Pistols..W. H. B. Smith
The Book of the Pistol & Revolver...........................Hugh B. C. Pollard

Made in United States
Cleveland, OH
05 November 2026

47764843R00166